the

FOREST
CERTIFICATION
HANDBOOK

Critical response to *The Forest Certification Handbook*
by Christopher Upton and Steve Bass

'The new Intergovernmental Panel on Forests will be addressing the transition to sustainable forest management. One of the key issues for its deliberation is the possible scope for forest certification as a means for opening up markets to the products of sustainably-managed forests. This book is timely, for it addresses certification in a balanced and technical way. It should be a very helpful input into the decisions that need to be made at the international level, as well as within individual nations and enterprises.'

Sir Martin Holdgate, former Director-General, IUCN

'An interesting overview of present discussions on a very complex subject.'

Claes Hall, International Development Director, Aracruz Celulose SA

'This is a clear and balanced overview of the key issues relating to timber certification. I strongly recommend it.'

Chris Elliot, Senior Forests Officer, WWF International

'Explains clearly how a certification programme should be run, discusses critically what certification may or may not achieve in terms of solving the problems that beset forests, and gives an up to date account of the relationship between certification and the other efforts that are being made to improve forest management worldwide. I strongly commend it to readers whether they be forest managers, traders, government officials or simply those who are concerned for the future of forests.'

Duncan Poore, from the Foreword

the FOREST CERTIFICATION HANDBOOK

Christopher Upton and Stephen Bass

CRC Press
Taylor & Francis Group
Boca Raton London New York

CRC Press is an imprint of the
Taylor & Francis Group, an **informa** business

First published 1995 by Earthscan Publications Limited

Published 2019 by CRC Press
Taylor & Francis Group
6000 Broken Sound Parkway NW, Suite 300
Boca Raton, FL 33487-2742

A catalogue record for this book is available from the British Library

Typesetting and figures by PCS Mapping & DTP, Newcastle upon Tyne

Cover design by Elaine Marriott

**Visit the Taylor & Francis Web site at
http://www.taylorandfrancis.com**

**and the CRC Press Web site at
http://www.crcpress.com**

Contents

Part 3 Current Initiatives and Views

Annexes

Part 4 Directories

List of Illustrations

Figures

Boxes

Tables

Foreword

This book is about the certification of forests - about providing a guarantee to the buyer that his or her purchase comes from a well-managed forest and will not favour unsustainable or inequitable practices. Such a guarantee should bring several advantages. First, it should satisfy the conscience of the buyer, and secondly, it should give added confidence to the various stakeholders in the chain of production (forest owner, forest manager, harvester, shipper, distributor and retailer) that the product will have a market and perhaps even fetch a better price.

But more important is the role of certification in providing an incentive to forest owners and managers to raise the standard of their forest management. It is, of course, not the only force in this direction. A new set of international norms was set when nations agreed to the Forest Principles at the United Nations Conference on Environment and Development in Rio de Janeiro in 1992. As a direct result, there are now many international and national initiatives to establish more specific principles for forest management, and to identify criteria to judge the quality of management, and indicators to assess performance. These initiatives address standards at the country level; but implementation will, of course, have to be at the level of the individual forest - the forest management unit. These two efforts - moves towards certification and efforts to define standards - are complementary; as they develop they should converge and become mutually supportive.

In fact, a great deal is going on. The whole field is developing rapidly and the links between the various initiatives are not always clear. Indeed, the picture presented to the observer from outside is often confusing. Sometimes it seems not altogether clear to the practitioners! Moreover this confusion could easily frustrate the excellent work that is now being carried out.

Here lies the importance of this book. It explains clearly how a certification programme should be run. It discusses critically what certification may or may not achieve in terms of solving the problems that beset forests. And it gives an up-to-date account of the relationship between certification and the other efforts that are being made to improve forest management world-wide. I strongly commend it to readers, whether they be forest managers, traders, government officials or simply those who are concerned for the future of forests.

Duncan Poore
Glenmoriston
August 1995

Preface

In this book we offer a discussion of forest certification for forest managers, commerce and industry and those involved in making policy. The book's layout is intended to give practical advice on developing, selecting and operating a certification programme.

One of our main concerns is that it should be clear what certification can and cannot do. Hence we examine the assumptions that lie behind forest certification; we present the use and limitations of certification as a tool which can help to solve forest problems; and, where certification does have a clearly defined role, we show how certification programmes should be structured and operated. This last point covers the issues which need to be addressed at the forest level by the certification body and the forest manager.

The Introduction shows the context in which certification has developed, and sets out the key definitions which are used in this book. Here we also set out the limits of what is covered and specify the book's objectives.

Part 1 goes on to describe in more detail the environment in which certification must operate. We start, in Chapter 2, by summarizing forest problems and their causes; and show, in Chapter 3, that different stakeholders are both subject to different problems, and hold their own perspectives about what is important. Here we introduce sustainable forest management as a goal where different perspectives and needs require reconciliation. We believe that definite principles can be established; but that to translate this abstract into practicable standards for forest management requires consensus between stakeholders - often with many differing views. This is an emotionally charged debate and practical solutions will, of necessity, evolve over time. Meanwhile forest managers need to understand and accommodate the views of stakeholders; and to establish mechanisms which allow these views to be incorporated in their activities.

At the forest level, practicable solutions are required. But to generate and support forest-level solutions, there are country- and global-level needs which require addressing; some of which are important to the success of certification. In particular, activities at the forest level are affected by the country forest policy environment and certain international commitments. In Chapter 4, we explore the policy context which supports certification, and discuss some of the more important incentives which need to be in place.

The process of achieving consensus on sustainability at the forest level inevitably hinges upon the issue of standards. Certification requires standards against which the performance of forest management can be assessed. It is our view that this is a debate in which there is much confusion. In particular, we find that general principles and criteria are muddled together with detailed standards which ought to be specific to the local forest management unit. Chapter 5 clarifies what the standards hierarchy should look like and how standards should be developed. We also examine many of the issues which have appeared in current attempts at standard-setting. A key point, which is often forgotten, is that adequate interpretation of standards needs to be done by *both* the forest manager and the certification body.

Building on the context set out in earlier chapters, Chapter 6 moves on to identify where certification fits. In this chapter we focus firstly on the use of and benefits from certification as a market-based instrument; and secondly those forest problems which certification helps to address.

Part 2 concentrates on the practical aspects of certification. Chapter 7 looks at design issues, describing and recommending the type of structure that a certification programme should have. This chapter demonstrates how local forest management units should be organized; how certification bodies should be structured and operate; and the importance of accreditation both in giving certification bodies a credible mandate and in attesting to the quality of their judgement.

Chapter 8 takes the reader through the actual process of certification. We describe the activities to be undertaken by the forest manager during the certification process and those to be done by the certification body. Each of the steps involved is set out in considerable detail to provide those considering certification, either as forest managers or as policy-makers, with clear details of what is involved. At the end of this chapter we include case studies which describe the experience of certification in four different forest situations.

The final part of the book, Part 3, describes current initiatives which either directly concern certification or have an impact on it [Chapter 9]. Lest it be thought that there is consensus over the merits and operation of certification, Chapter 10 shows what some of the current views on certification are. There is obviously certain confusion and difference of opinion. The debate becomes particularly lively over the role that governments could play in certification. In the light of the preceding chapters, suggestions are made regarding the next steps at local, national and international level.

Christopher Upton,
SGS Forestry

Stephen Bass,
International Institute for
Environment and Development

Acknowledgements

Many people have contributed to the realization of this book and have very helpfully provided information and time. We would like to thank: Helena Alburquerque, Rainforest Alliance, USA; Vittorio Alpi, Alpi Spa, Italy; Paul Ankrah, Do-It-All, UK; Peter Barden, SGS Forestry; Ronnie di Camino; Rosalind Reeve, Pro Regenwald, Germany.Marcus Colchester, World Rainforest Movement; J Clarette and S Caron, Canadian Forest Service; Janet Croucher, Demerara Timber Ltd, Guyana; Chris Elliott, WWF International; Michael Groves, SGS Forestry; Claes Hall, Aracruz, Brazil; Joseph Häckl, Umweltbundesamt (Federal Environment Agency), Austria; Debbie Hammel, Scientific Certification Systems, USA; Pierre Hauselmann, consultant, Switzerland; Sir Martin Holdgate, UK; Dorothy Jackson, Soil Association; Jean-Paul Jeanrenaud, WWF International; Klaas Kuperus, Vereniging Van Nederlandse Houtondernemingen, Netherlands; Hubert Kwistout, Ecological Trading Company, UK; Anders Lindhe, WWF Sweden; James Lonsdale, Silvanus, UK; James Mayers, IIED; Frank Miller, SGS Forestry; Elaine Morrison, IIED; John Palmer, Tropical Forestry and Computing, UK; Duncan Poore, UK; Ravi Prabhu, CIFOR, Indonesia; Andrew Roby, Natural Resources Institute, UK; Per Rosenburg, WWF Switzerland; Sari Sahlberg, Ministry of Agriculture and Forestry, Finland; B Saua, Soltrust, Solomon Islands; Mark Smith, SGS Forestry; Jonathan Sinclair-Wilson, Earthscan; Hugh Speechly, SGS Forestry; Francis Sullivan, WWF UK; Ian Symons, UK Government;

Helpful research was provided by Janet Probyn (independent researcher) and Rebecca Munden (SGS Forestry).

1. Introduction

Concern for forest problems has increased dramatically over the last decade. For numerous reasons over 200 million hectares of forest have been lost in the tropics and large areas of boreal, temperate and tropical forest have degraded in quality. As pressures have increased on remaining forest areas, conflicts have grown between 'stakeholders' – those who live in forests, forest industries, governments, and the public at large – who depend in different ways on the environmental, social and economic benefits provided by forests.

The traditional – usually government-led – approach to forest problems has been regulatory. In poor countries this approach has often been supplemented by aid-funded programmes. In general, these efforts have proved insufficient to reduce either forest loss or forest degradation. At the country level, forest legislation may be inadequate to assure improvements in forest management; and customary rules governing local forest use may not be recognized. Alternatives are required to redress the deficiencies in existing mechanisms. There is a need to recognize the wider asset value of forests throughout the world; and for new instruments to be developed which enable forest owners in rich and poor countries to get the best return within a context of sustainable forest management.

In the meantime public impatience – especially in North America, Europe and Australasia – with lack of progress and disillusionment over the effectiveness of existing forest initiatives are resulting in moves to look at the possibilities of market-based, voluntary approaches. The assumption behind these initiatives is that consumer interest in the forest dilemma is strong.

It is further assumed that this interest may cause discrimination in favour of timber from sustainably-managed forests, and a willingness to pay any associated extra cost. It is also thought that public acceptability of wood and paper products from sustainably-managed forests will help to maintain their market share against substitute non-wood products. This is based on the assumption that the public appreciates the inherent virtues of wood and paper products as deriving from a renewable resource and being ultimately biodegradable.

However, the converse of this assumption worries some stakeholders – namely that consumer concern over forest conditions may result in a discrimination against timber and paper products that the consumer *perceives* to derive from unsustainably-managed forests.

These assumptions have provided the impetus for development of forest certification – hereafter referred to as certification. This has four key parameters:

1   Certification has the twin objectives of (a) working as a market incentive to improve forest management; and (b) improving market access and share for the products of such management.
2   Certification is conceived as an economic, market-based instrument and as such participation in certification programmes should be, and currently is, voluntary.
3   Certification takes place by assessing the effect of forest activities against standards previously agreed as significant and acceptable to stakeholders.
4   Certification is undertaken by third party organizations which have no self interest in a specific forest activity; which are not stakeholders in the forests being certified; and which can assure the public of independent and professional judgement.

Governments, corporations and other people deriving their livelihood from forests, as well as consumers, need to be in a position to know whether the attention given to certification is justified. The key questions amount to:

•   Can certification and current certification initiatives be of real use in slowing forest loss and degradation, whilst assuring the objectives of improved forest management?
•   Is certification in fact a distraction from fundamental issues; or even actively damaging by pulling resources into activities that fail to create real change?
•   Is certification, as currently established, practicable? Will it provide participants with the commercial advantage they seek? Which groups might lose out if certification becomes widespread?

Current certification initiatives are developing against a rapidly-changing background of international and national initiatives in forestry, biodiversity, conservation, environmental management systems and trade – many of which also aim to achieve sustainable forestry on the ground. Within this context, there is scope to focus certification so that its potential contribution can be fully realized.

Sustainable forest management is an inherent aim of certification. It is the ultimate goal to which certified forests should aspire, but such a goal is reached only through a period of transition, during which management standards are progressively established and fine-tuned. The explicit aim of certification is to improve the quality of forest management so as to reach this goal.

The practice of certification has to be precise, unambiguous and repeatable in its assessment; while the paradox is that perspectives on sustainable forest management differ, and have often been expressed vaguely and ambiguously. Reconciling this difference is perhaps the most difficult part in developing and operating a certification programme.

Attempts at defining what is required in terms of sustainable forest management are almost always controversial. While it is possible to secure a consensus over general principles, it is difficult to achieve this for standards designed to be used at the forest level. In certification, we prefer to use the term quality forestry to describe a performance of forest management that is considered adequate: ie, that is basic to the transition to sustainable forest management.

Quality forestry is defined as that which is *environmentally sensitive, socially aware* and *economically viable*:*

- Forestry is defined as being *environmentally sensitive* when its impact upon the environment is both assessed and negative aspects minimized.
- Forestry is defined as being *socially aware* when it recognizes that its activities have an impact both on local people and society at large. Socially aware forestry, therefore, encourages all stakeholders to enjoy long-term benefits of the forest, provides strong incentives to local people to sustain the forest resources and adhere to long-term management plans, and provides for an appropriate distribution of costs.
- Forestry is defined as being *economically viable* when forest operations are structured and managed to be sufficiently profitable for ensuring the stability of operations and a genuine commitment to principles of quality forestry. Economic viability does not mean financial profit at the expense of the forest resource, the ecosystem, affected communities, or society as a whole.

The focus of a certification programme is at the forest level. In certification, this area is defined as the Local Forest Management Unit [LFMU]. Usually the LFMU is managed by a single owner or operator. However, there are instances where this is not the case. The area being certified must have a common management system but may be owned by several individuals grouped together, by a community or even lie within a single administrative boundary within which forestry activities are regulated by a national forest service. In this book, we refer to the area being certified as the LFMU. Selection of an appropriate LFMU is a critical part of the certification process having a significant impact on the costs of certification. Chapter 8 includes a discussion on selection of the LFMU.

In addition, certification will always be of a forest area that is being managed. Certification, therefore, assumes that prior decisions have been made with regard to different land uses and in setting aside areas as national parks for pro-

---

* This definition of quality forestry is consistent with the FSC principles used for certification – see Box 24 and the entry for FSC in Chapter 11 for further details.

tection and conservation. Certification cannot assess whether or not a particular forest area should be managed. In this book we assume that certification takes place in production forest areas designated at the national policy level as part of a permanent forest estate (PFE). Moreover in the future it is likely that acceptable mechanisms will be introduced to allow for conversion forests in certification – provided that such areas form part of an overall sustainable land use plan.

There is much confusion over the practice of certification at the national level. Certification requires an adequate policy context and certain incentives to be in place for it to be effective. These are best defined at the national level. Accreditation authorities may also be national bodies. However, in order for a certification programme to accommodate the needs of international trade, general principles and criteria should be established at an international level and interpreted for direct application at the forest level by the LFMU and certification body. Such an approach ensures international harmonization; it is the principle behind the development and certification of ISO standards – such as ISO 9000 for Quality Assurance; and is the one used in this book.

Sustainable development – the balanced achievement of economic, social and environmental objectives now and for the future – rarely has a single solution. This is because decisions on the relative values of economic, social and environmental factors are subjective; and consensus is nearly always required between stakeholders. No one stakeholder view has universal authority and all solutions need to be adaptable to local circumstances. If there is one fundamental premise of this book, it is that the role and practice of certification within this process of integration and trade-off has to be clearly identified; not only in terms of national forest and land use policy but also so that it fits with local needs and practicalities.

Part 1

Certification in Context

---

2. Forest Problems

A major reason for writing this handbook is to explore how certification could help to solve the real problems affecting forests and people dependent upon them. This chapter sets the scene by introducing the main forest problems and their causes.

Each country and each LFMU will experience a different set of problems to varying degrees. These problems mean that forests are unable to realize their potential contribution to economic and social development. Some of the problems have a global dimension as well. Broadly speaking, the most significant are:

- **Reduction of forest area and quality:** the quantity and quality of forests is declining. This is because wood, fuel, food and fodder are being cut at rates which are faster than forest regeneration; because remaining growing stock is often poorly managed; and because many forests are being cleared to make way for other land uses. Reforestation is not of equal quantity nor quality – that is, it does not replace all the benefits of natural forest.
- **Environmental degradation of forest areas:** forest exploitation and clearance can create other interlinked problems; notably soil erosion, watershed destabilization and micro-climatic change. Industrial air pollution, particularly common in some temperate forests, reduces forest health. Many forest environmental benefits cannot be supplied by other land users.
- **Loss of biodiversity:** the above problems are contributing to a rapid reduction in ecosystem, species and genetic diversity in both natural and planted forests. This lowers the world's biological potential for improving material, food and medicine production. With tropical forests being perhaps the major repository of biodiversity, forest abuse in tropical regions has caused much concern.
- **Loss of cultural assets and knowledge:** the often-undocumented culture and knowledge of many peoples, which have evolved through long

periods of nurturing the forest, are diminishing as forest area reduces, as access to forest is increasingly restricted, and as traditional rights are eroded. This leaves mankind as a whole with a smaller knowledge base for forest stewardship.

• **Loss of livelihood:** all the above problems are affecting the livelihoods of forest-dependent peoples – particularly poorer groups in poor countries who may not have significant agricultural land, and who depend on forests for 'social security'. With such people marginalized from the forest, social and economic problems are created elsewhere, such as in cities.

• **Climate change:** it is probable that the cumulative effect of global forest loss and environmental degradation will contribute to regional and global climate imbalances. Forests play a major role in carbon storage: with their removal, carbon dioxide in the atmosphere may lead to global warming with its many problematic side-effects.

Although each stakeholder group within a country tends to emphasize only one or two problems (the environment, or land rights, or poverty etc), the more the problems are analysed, the clearer it becomes that these problems are linked. Forest problems are the result of a syndrome of many causes; and action on only one front will rarely solve them.

Many of the causes which underlie most forest problems arise outside the forestry and forest industry sectors. Consequently, activities from within these sectors alone are unlikely to solve forest problems.

Basic market, policy and institutional failures tend to either 'push' groups into the forest, through marginalizing them in places outside the forest; or to 'pull' groups into the forest, through attracting them into the forest by excess profits. Many of the policy failures concern agriculture and industrial development, or are a result of inadequate macro-economic policies. The effect of these failures can be worsened by certain demographic, physical and technological conditions. This subject is an enormous one, of which only some basic points are laid out below. Naturally, not all of the causes cited apply in any one country or circumstance.

Market and policy failures within and outside the forestry sector

These have a number of effects:

• they undervalue forests, eg low stumpage fees for timber;
• they overvalue the benefits of removing forests, eg subsidised agricultural prices;

- they do not reflect the social and environmental externalities[1] of forest management or removal, eg by not requiring that the operator covers such costs; neither are these externalities included in measures of national economic performance;
- they make investments in sustainable forest management unprofitable or risky, eg high interest rates and lack of long-term financial stability; and
- they count against primary production. Producers of raw materials such as wood have seen their long-run terms of trade deteriorate in comparison to manufactured goods. Primary producers tend to receive a small proportion of the final product price, typically 10 per cent.

Institutional failures

Institutional failures can explain many of the root causes of forest problems, and are strongly related to policy failures. They can be grouped in the following way:

- Poor information and monitoring of forest stocks and flows, and of changes affecting forests. Forest managers often have poor information on forest potential and behaviour, making it difficult for practices to be both productive and sustainable.
- Differing priorities between stakeholders, together with a lack of participation and mechanisms by which consensus can be reached.
- Uncoordinated decision-making, which so often leads to conflicts between government policies, and other inadequate responses.
- Unclear or outdated institutional roles, both within the forestry sector, and between the forestry sector and other sectors, resulting in inefficient, incomplete, duplicating or conflicting work.
- Misdirected international assistance for poor countries.
- Government control mechanisms which are inappropriate, weak or ineffective.
- Inequity between and within nations, in access to the benefits of forests, and to the resources required for their management and use.
- Lack of political will, as influential stakeholders tend to be unwilling to reach compromises between their demands on forests and those of others.
- Covert institutional relationships, eg systems of patronage that affect outcomes in spite of formal policies.

---

1. 'Externalities', such as loss of biodiversity and social dislocation, are traditionally considered by economists as external to the economic system; their costs [or benefits] are not included in prices. For example, the cost of watershed protection would not be included in the price of wood extracted from such areas. Additional incentives, such as subsidies, are required to ensure appropriate forest management.

Weak and/or inappropriate tenure

Together with conflicts between land use policies and local rights, these lead often to forest problems concerning weaker groups. Notable causes are:

- tenure systems which require deforestation in order to obtain title, and which encourage speculation;
- poor recognition of access and user rights for the landless, people with traditional claims to forests and/or weaker social groups who may depend upon forest resources;
- forest tenure systems with inadequate security, discouraging settled, long-term forest activities;
- historical anomalies abounding in tenure systems – particularly concerning access and use rights, but not being removed if they suit local elites;
- governments enforcing extreme forms of tenure, such as nationalization or issuing of forest concessions which do not recognize traditional, mixed and/or overlapping forest access and use rights; and
- poor national records and demarcation of forest tenure, constraining initiatives which depend upon forest manager compliance.

In addition other – more direct – causes can exacerbate the failures of markets, policy, institutions and tenure:

- **Increasing population:** growing and migrating populations require land for settlement and food production; often forests appear to be the most freely available land.
- **Increasing demands for forest products:** greater wealth and larger populations require increased levels of harvesting from ever-smaller forest areas.
- **Fragmentation of the forest:** clearance and small-scale reforestation increases the physical challenge of protecting patchy forests and controlling their use; and disproportionately reduces their biodiversity and watershed capabilities.
- **Increasing extent of infrastructure:** roads and railways increase the possibilities for uncontrolled access to previously inaccessible forests.
- **Inappropriate technology and skills applied to forest management:** simple, mechanized approaches with high potential environmental and social impacts are dominant; and application of traditional multiple-use management is dying out.

In conclusion, there are almost always several causes of a specific forest problem, and these interact in complex and often unpredictable ways. Consequently it is not surprising that single-issue, single-stakeholder or single-tool solutions have failed to alleviate forest problems.

Although, as we shall see, certification is designed to address multiple issues and multiple stakeholders, it is still a single-tool solution whose limitations should be recognized. The main limitations of certification are: (a) it cannot address policy and institutional failures; and (b) it cannot directly improve land use decisions – it has to work within the scope of basic forest use and management decisions that have already been made. These limitations are a consequence of certification having to focus at the forest level. Hence certain policy requirements are, in effect, prerequisites for certification to reach its full potential. Between them, these policy requirements and certification need to provide the right conditions for the transition to sustainable forest management. These many policy needs are discussed in Chapter 4. The ultimate goal of sustainable forest management is the subject of the next chapter.

3. Competing Stakeholder Interests and the Goal of Sustainability

Competing stakeholder interests

The causes of forest problems are bound up with the diverging and competing interests of 'stakeholders' – such as government, communities in forested areas, employees, investors and insurers, customers and consumers, environmental interest groups and the general public. Stakeholders have differing specific interests in the goods and services provided by an LFMU.

Demands on forests from various stakeholders differ – although relationships between stakeholders and the strong influences of some groups rather than others do produce a degree of convergence. Stakeholders in different sectors – such as agriculture, tourism, energy and forestry – will also place varying demands and claims on forest goods and services, or forest land. Some stakeholders, such as environmental groups, are extremely effective at advocating certain perspectives. Incompatibility between stakeholder demands is often a result of the degree to which biodiversity conservation is considered essential to good forestry. Indeed, stakeholder differences are bound up in the ongoing debate aimed at achieving consensus on what is meant by sustainable forest management.

The mix of differing stakeholder demands, how they interact with attempts to define sustainable forest management, their impact upon the LFMU and certification as a possible tool available to the forest manager are shown in Figure 3.1. Certification may provide a mechanism for the LFMU to deal with stakeholder demands. However, in order to resolve conflicting perspectives, processes of mutual learning and consensus-building are also required. These are particularly important today and are emphasized in Chapter 4. They also take time.

One thing that unites almost all stakeholders, from those with local interests

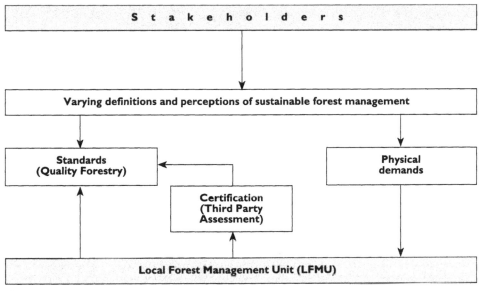

Notes:
1 The LFMU is under increasing pressure from stakeholders demanding improved forest management.
2 Stakeholders require that forest management become sustainable.
3 Stakeholders have varying perceptions of and definitions for sustainable forest management.
4 Because of these differences the forest manager requires clarification on what is meant by sustainable forest management.
5 The forest manager requires achievable standards, acceptable to stakeholders, to work to.
6 Standard setting requires consensus and trade-off between stakeholders; and is difficult as it revolves around a definition of sustainable forest management.
7 For this reason many standard-setting initiatives do not refer to sustainable forest management. In this book we use 'quality forestry' as an interim and achievable goal.
8 Stakeholders require credible proof that the LFMU has adopted the agreed standards before releasing pressure.
9 Certification is one mechanism where the forest manager can achieve this.

Figure 3.1 *Stakeholder pressures and sustainable forest management*

to those with global interests, is that the concern they are expressing for the state of forests is greater than at any other time. Despite this, a large proportion of stakeholders is often left out of the decision-making process. They are confronted with evidence showing continued forest degradation and loss – environmental interest groups have proven themselves to be effective at leading stakeholder opinions and forming new perceptions and demands, through both the market and public policy (see Box 3.1). Yet traditional systems that might have reconciled stakeholder demands have collapsed or have been dismantled by governments intent upon governmental control of most forest matters. Current political systems are often incapable of reconciling stakeholder demands; and because of the long time it takes to become enacted, legislation may not meet current needs. In addition, the economic and technological power of individual stakeholder groups is now capable of causing more rapid change. As a result, many stakeholders feel alienated and impotent, and governments are not in an immediate position to help.

Box 3.1 *Environmental groups – impacts on other stakeholders*

Campaigning environmental pressure groups: the direct and indirect effect that the activities of these groups have had on stakeholders is significant. As Figure 8.6 (page 106) suggests, the activities of such groups are arguably behind the strongest external pressures facing most forestry organizations. Often the pressure is indirect, as environmental interest groups influence other agents more directly connected to the forestry organisation.

Campaigning groups have a wide variety of both agendas and methods. Such groups are, however, united by a common thread which is to improve, through public pressure, environmental performance and responsibility. By their nature, these groups also have a range of thresholds at which they will accept that their objectives are fulfilled. Some of these can be accommodated by LFMUs and some of them cannot.

Environmental interest groups as partners: other environmental groups which are equally effective in maintaining external pressure prefer dialogue and partnership to confrontation. Such groups tend to work with a broad range of stakeholders – from governments to communities and the private sector. Some work with LFMUs, hoping to influence their environmental performance as models and examples. Some are important in influencing regulators and legislators with regard to improvements in policy etc. They are particularly important as catalysts of sustainable development, drawing in the various stakeholders, providing cross-sectoral information and aiming to reach a consensus on trade-offs. Their work is often backed up with detailed research.

Many environmental groups operate in ways which fall between these two types.

The current situation of unfulfilled legitimate stakeholder demands and inappropriate response is illustrated in Box 3.2 with regard to Cameroon. In Cameroon the forestry sector is currently experiencing change. New legislation is being enacted in order to restore past imbalances; and ground demarcation of the permanent forest estate in the south and east of the country will adopt a participatory approach once sufficient funds are available.

It is the LFMU which receives the cumulative impact of stakeholder pressure, illustrated in Figure 3.2. Stakeholder pressures are often inter-related. For example, increased regulatory pressure may be a response to local community concern and campaigning by environmental groups. A general increase in environmental pressure may cause financial markets, lenders and insurers to perceive increased environmental risk and so exert pressure themselves. This can occur even in the absence of 'scientific' proof. For example, global insurance markets have decided to require insured parties in many coastal areas to take action against prospective sea level rise – an environmental phenomenon about which scientists are not yet agreed. Some stakeholder perceptions can therefore be very significant.

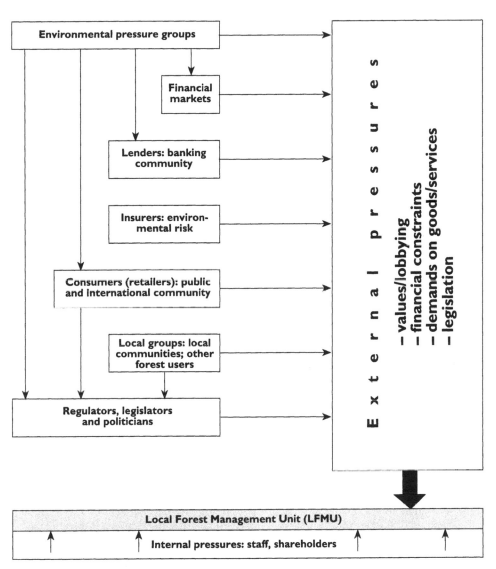

Figure 3.2 *External pressures on a forestry organization*

Box 3.2 *Different stakeholder perspectives on forest issues: an example from Cameroon*

Cameroon is one of the most important timber exporters in Africa. The country is covered by 20 million hectares of tropical humid forest, which is about 40 per cent of total land area. The annual harvest of wood is 2.5 million m^3 per year, of which over 1 million m^3 is exported in log form. Forest exploitation is increasingly motivated by the need to collect government revenue. Such national policy needs may clash with those of other stakeholders, at global and local levels.

From an international perspective Cameroon is one of the most ecologically diverse countries in Africa. Many international conservation groups, such as WWF, are calling for increased conservation. Yet logging in protected areas is common, as there is little coordination between conservation and logging interests. The result is that, in many instances, forest exploitation has managed to take precedence over ecosystem protection.

Local people may not be primarily concerned with the impacts of logging activities and the environmental consequences. They acknowledge these and suffer from some of them; but they also welcome the access to markets that logging roads provide and the schools and clinics which are built. They may, however, have different long-term interests in the forest not shared by other stakeholders. For example, the Moabi tree provides food and oil from the seeds and medicine from the bark. It also ranks as an important timber species and logging has reduced populations in many areas.

Local people dependent on the forest have traditionally been excluded from the policy-making process and the planning of forest exploitation. The existing laws and policies focus on relations between the state and the logging companies. Local people feel that official policy on the forest does not take into account their interests, even though the availability of bush meat, medicinal plants and other wild foods is directly affected by logging activities.

Forest-related donor initiatives have led to the commissioning of numerous studies and the proliferation of activities with little regard for their effectiveness on the ground. One aspect of donor involvement is a lack of coordination between different initiatives. Such initiatives are often imposed and do not always have the active support of the government nor of local people. Some initiatives are influential at policy level, but others serve to reinforce logging activities; or are preoccupied with 'global' diversity.

Sustainable forest management is constrained through lack of an appropriate institutional and legal framework. The participation of government, local people, environmental groups and the forestry companies is required to ensure the long-term ability of the forest to produce the goods and services required. These groups need to work together to define, use and continually refine forest usage and responsibilities. The role of donors should be very much in the background, as a support to this process and to help build stakeholders' capacities.

New legislation and demarcation of a permanent forest estate is being prepared in Cameroon, to take into account all forest goods and services. It aims to arrive at a consensus between stakeholders. The same principle should apply to certification and standard setting.

As the general forest environment continues to deteriorate, the external pressures exerted by stakeholders on the LFMU are gradually increasing everywhere. How much pressure is felt by individual forestry organizations depends on three main factors:

- **Source of wood:** whether directly operating in, or procuring wood from, natural forest anywhere in the world, but especially the tropics and in forest perceived as 'old growth'.
- **Organizational size and whether listed on the stock exchange:** and therefore relative visibility and financial importance of the organization.
- **Location of head office and operations:** and therefore whether easily targeted by campaigning environmental groups.

LFMUs which fit all three of these criteria tend to be under considerable pressure for accountability with regard to their forestry operations. This is true even for organizations which may not even have forestry operations of their own, but which manufacture and/or sell wood and paper products.

In particularly sensitive forest areas, it must seem to the forest managers on the ground – who often are just trying to do their job – that the world is against them. The LFMU appears to face a cascade of pressures. Few fully understand the external environmental pressures which they face; and the exact implications for their operations. Despite this situation, only a few forest managers are already making decisions to improve their performance in light of broader demands for sustainability.

Eventually public and scientific opinion, if strong enough, will result in legislative changes required which will force forest organizations to act differently. In the meantime, organizations adopting a *reactive* strategy may suffer from poor environmental reputations and a poor skills base to deal with environmental issues when they do require attention. This approach may affect their position in a competitive marketplace.

Proactive organizations, which aim to ensure that environmental management anticipates future legislation and is in harmony with current and likely stakeholder demands, are more likely to gain competitive advantage from the situation. Being proactive often means being visible, so that it is important to be *successfully* proactive. A proactive strategy requires skills, commitment and discipline for it to work.

The external pressures faced by LFMUs are, to a certain extent, complemented by internal pressures from staff and shareholders or owners. An LFMU's ability to attract well-qualified and motivated staff is almost certainly affected by its 'environmental' reputation. Most forestry organizations are currently struggling to accommodate the new stakeholder pressures on forests. Some of those who are struggling, fail to understand and/or respond to the needs of the external environment which they face. There is considerable mistrust of other stakeholders and misunderstanding concerning the practical implications of stakeholder perceptions and aspirations. This is partly because some stakeholder groups may not be entirely clear about their requirements.

Certification in Context

It is for these reasons that stakeholder demands need to be clearly set out and consensus reached concerning the rules under which LFMUs can operate. Achieving this consensus is inextricably bound up with debate over the goal of sustainable forest management and the standards to which certification should operate.

The goal of sustainable forest management

The debate on what constitutes sustainable forest management is evolving at the same time as initiatives on certification. Certification is designed, ultimately, to lead to sustainable forest management. As an interim goal, we propose the notion of *quality forestry* as that which accommodates a level of forest management which is acceptable to stakeholders (see Chapter 1). Nevertheless, with regard to sustainable forest management, it seems that there are certain general principles of sustainability which have been agreed. These include:

- **Environmental sustainability:** this entails an ecosystem being able to support healthy organisms, whilst maintaining its productivity, adaptability and capability for renewal; it requires that forest management respects, and builds on, natural processes;
- **Social sustainability:** this reflects the relationship between development and social norms: an activity is socially sustainable if it conforms with social norms, or does not stretch them beyond a community's tolerance for change; and
- **Economic sustainability:** this requires that benefits to the group(s) in question exceed the costs incurred, and that some form of equivalent capital is handed down from one generation to the next.

In making the transition to sustainable forest management in any given locality, the following aspects also need to be included:

- achieving *environmental objectives*, eg maintaining biodiversity, watershed quality and climate regulation;
- achieving *economic objectives*, eg maintaining timber yield and the forest's capital value;
- achieving *social objectives*, eg meeting the livelihood needs and maintaining the cultures and knowledge systems of forest-dependent people;
- *balancing today's needs* with those of future generations;
- *integrating the above objectives* where possible – exploiting the 'win-win' possibilities;
- making informed *trade-offs between objectives* where integration is not possible – cost-benefit analysis and environmental impact assessment can help to achieve maximum positive impacts and minimum negative impacts, both on-site and off-site;

- balancing the *different levels* at which needs and demands are expressed, from the forest compartment, to neighbouring communities and land uses, to the nation, to the globe – at each level, the integration and trade-off possibilities will be different, and consensus-building and negotiation can help;
- allowing for *complexities*, and many *uncertainties* – we will not always know the possible outcomes;
- adopting procedures for *continuous improvement*, which requires an emphasis on monitoring and learning;
- ensuring *participation of all stakeholder groups in decision-making*, and transparency of decisions;
- providing good *information access* for all stakeholder groups with legitimate interests;
- gaining the *commitment of the 'losers'* of each decision, as well as the 'winners';
- *ensuring stakeholders are held accountable* for their forest activities, given that these will have impacts on other stakeholders – this may require independent verification, such as certification may provide; and
- *long-term, policy-level support* and financial stability for sustainable forest management.

Sustainability depends upon the specific relationships of forest management with the surrounding environment and society. Few criteria of sustainability will be universal. Precise definitions of sustainability need to be locally negotiated, and revised as technology and society's demands, and other conditions, change. Hence the importance of management plans, participation of stakeholder groups, monitoring, continuous improvement and a cyclical approach to policy-making.

The relative importance of effects and objectives will always differ according to the situation. For example, achieving sustainability for *plantations* may require a focus on impacts on soil and water regimes, work safety and employment; and equity impacts in terms of changed access to food, water and firewood and impacts on incomes. A sustainable plantation would be likely to adopt priorities based on:

- profitable and efficient production, with an emphasis on improved whole-forest management that is site-specific; and which uses a systems approach to conserve soil, water, energy and biological resources;
- a more thorough incorporation of natural processes such as nutrient cycles, nitrogen fixation, and pest-predator relationships into the forest production process;
- a reduction in the use of external inputs with the greatest potential to harm the environment or the health of forest operations, and a more targeted and efficient use of the remaining inputs used;
- a greater productive use of the biological and genetic potential of timber species and animal species; and

- an improvement in the match between tree species and the productive potential and physical limitations of forested lands, to ensure the long-term sustainability of current production levels.

In contrast, sustainability of *tropical rainforest* management would require additional emphasis on significant biodiversity, customary access and the availability of non-wood forest products.

It is perhaps clear from the above that achieving the goal of sustainable forest management is a multi-dimensional balancing act. In today's world, new procedures and initiatives are required in order to support the process of achieving sustainable forest management. Certification, as we shall see, has a clear potential. However, effective policy formulation and implementation will be important prerequisites, and are the subject of the next chapter.

4. Policy Requirements

We have described how forest problems are due to: policy, market and institutional failures; inadequate tenure; rising populations and their demands; fragmentation of the forest estate; and inappropriate infrastructure, technology and skills. The transition to sustainable forest management primarily involves tackling these problems.

This requires policy decisions to be made at national and international levels; the subject of this chapter. Decisions which are required at the local level are discussed in the appropriate sections of later chapters; see especially Chapter 5 on local level standards and Chapter 7 regarding management systems.

National level

At the national level, basic policy initiatives are required to make the transition to sustainable forest management. They need to tackle the many forest problems that have their roots in perverse or conflicting legislation and regulations, and to establish incentives for different stakeholders. Before looking in more detail at these, we should stress that the *way* in which policy decisions are reached is also important. A *strategic, participatory approach to national forest policy, emphasizing continuous improvement over time* is required. A considerable body of evidence from tropical forestry action plans, to national conservation strategies, to green plans etc shows this to be essential [Carew-Reid et al 1994]. The transition to sustainability will require several 'turns' of a cycle of goal-setting, planning and capacity-building, field management, monitoring, information assessment and goal-revision – an approach which is illustrated in Figure 4.1.

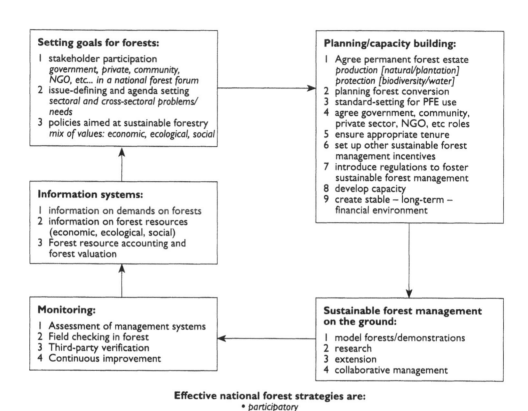

Setting goals for forests:

1 stakeholder participation
  *government, private, community,
  NGO, etc... in a national forest forum*
2 issue-defining and agenda setting
  *sectoral and cross-sectoral problems/
  needs*
3 policies aimed at sustainable forestry
  mix of values: economic, ecological, social

Planning/capacity building:

1 Agree permanent forest estate
  *production [natural/plantation]
  protection [biodiversity/water]*
2 planning forest conversion
3 standard-setting for PFE use
4 agree government, community,
  private sector, NGO, etc roles
5 ensure appropriate tenure
6 set up other sustainable forest
  management incentives
7 introduce regulations to foster
  sustainable forest management
8 develop capacity
9 create stable – long-term –
  financial environment

Information systems:

1 information on demands on forests
2 information on forest resources
  (economic, ecological, social)
3 Forest resource accounting and
  forest valuation

Monitoring:

1 Assessment of management systems
2 Field checking in forest
3 Third-party verification
4 Continuous improvement

**Sustainable forest management
on the ground:**

1 model forests/demonstrations
2 research
3 extension
4 collaborative management

Effective national forest strategies are:
• *participatory*
• *cyclical and continuously improving*

Figure 4.1 *A strategy for sustainable forestry at the national level*

Establishing multi-stakeholder involvement in decisions on forests

Decisions on how to deal with forest problems should be *cross-sectoral*, recognizing the varied causes and impacts of forest problems. The approach should also be *participatory*, with mechanisms to bring stakeholders together, with equal access to information, in a continuing process aimed at building consensus. In other words, participation is required throughout the cycle illustrated in Figure 3.1, although it will take different forms for each task.

This process should have *high-level support* over the long term, so that fundamental policy and institutional improvements have a chance to take place.

The first step is to seek various perspectives on forest issues, and then to set the agenda for discussion. It will not always be easy to reach consensus on the agenda – let alone the goals that should be set. For example, the debate over definitions of sustainable forest management has become very confrontational and controversial. Equally confrontational and controversial is the extent to which current forest practices have been said to match any definition of sustainable forest management. As current forest practices often represent the level of forest management required by law, casting doubt on these practices calls into question both the efficacy of the law itself and the public bodies responsible for its implementation. Some stakeholders then perceive forest users and regulators as suspect in their ability to guarantee sustainable forest management. This tends to drive stakeholders apart, and it is difficult to achieve consensus between them. Total consensus is therefore probably impossible, at least initially. The best that can be achieved is the accommodation of most of what most stakeholder groups are aiming for. If an iterative approach is taken, then stakeholder needs may, over time, be more closely met.

Appropriate policy and legislation

Investment in sustainable forestry is necessarily long-term. LFMU managers require that market, fiscal, tenurial and legislative conditions are clear and consistent over long periods of time, if they are to have confidence in their ability to reap the rewards of investment in quality forestry.

Forest policy and legislation also need to be increasingly well harmonized with those of other sectors, and aimed at generating the stable land use patterns that will contribute to sustainable development.

Agreeing on, setting up, and managing a Permanent Forest Estate (PFE)

A fundamental policy decision for sustainable forest management faced by governments, is to set aside a national Permanent Forest Estate (PFE). Until a PFE is set up, there can be very little basis at the national level for declaring that forest management is sustainable: at any time, large areas of forest can be vulnerable to clearance. The PFE should cover legal classifications of production forest (natural and plantation); protection forest (for biodiversity, cultural and watershed conservation) and mixed land use categories.

The PFE and its ground demarcation need to be agreed by all stakeholders; it is really a question of 'how much forest do we need, now and for the future, and taking into account all the different local, national and legitimate global demands'. Major issues to be resolved include:

- how large the national PFE should be to provide the goods and services required;
- the balance required between production and protection forests;
- the extent to which plantations can substitute for natural forests within the PFE;
- how much forests outside government ownership should contribute; and
- the rational planning and use of 'conversion' forests to be excluded from the PFE and turned towards non-forest uses.

Certification in Context

19

It should be noted that PFEs are already incorporated in the land use planning activities of many countries. Particularly in the tropics there is, however, an urgent need for effective ground demarcation of planned PFEs.

Select policy instruments which are effective incentives for sustainable forest management

The supply of products and services from forests needs to be kept within sustainable limits. This means that the signals received by forest managers must encourage them to manage and use forest resources sustainably. This, in turn, requires a mix of economic, institutional and regulatory instruments which reflect the full value of forests, as well as the social and environmental costs of forest use. They must be targeted closely on the specific stakeholder group; and hence must be designed in full knowledge of stakeholder behaviour and motivations. A review of some instruments available is given in Table 4.1. More specific information on certification as a market-based instrument is offered in Chapter 6.

Secure tenure and rights over forest resources

Tenure defines who has access to what parcels of land and/or resources, for how long, and for what purposes. The issue of *who* owns the forests, *who* occupies them and *who* makes what use of them, is extremely significant.

Conscious attention to tenure must be made, to ensure that land and forest use patterns are productive, sustainable and equitable. Tenurial changes may be necessary to ensure that:

- occupiers of forest land have adequate, stable, clear and secure tenure in order to use forests sustainably;
- no perverse tenurial incentives exist to deforest the PFE;
- there are adequate incentives to afforest land; and
- formal tenure recognizes and secures equitable rights for landless and/or weaker social groups who may depend to a great extent upon forest resources.

Define more appropriate roles for stakeholders

The roles and mandates of the different groups having an impact on forests need to be agreed and clearly defined. They also need to be realistic. This process may require changes to existing organizational structures; it has frequently resulted in the devolution and/or privatization of central government functions.

Often, within *government*, a rationalization and better coordination of land use activities is required. Governments need to be able to exercise adequate authority and control over those forests which have the highest public value, such as for the conservation of biodiversity, the protection of watersheds and the maintenance of recreational access. Here, private or community roles alone would be inadequate.

Table 4.1 *Examples of instruments available for the integration of policy for sustainable forestry*

Economic approaches	Institutional and advisory approaches	Regulatory approaches
• input charges and taxes as part of Polluter Pays Principle • remove or restructure existing production subsidies and taxes • implement subsidies for environmental goods and conservation • encourage cultivation of non-wood forest products in forest areas • transferable harvesting permits and quotas	• direct advice to forest users • education via media to forest users and other stakeholders • reform of sectoral agencies and modes of work of professionals • incentives to encourage research into sustainable forestry • better consumer information and choice and marketing: *certification* • formation of local groups for participation and empowerment • more private and public partnerships	• restrictions on damaging practices • prohibition of undesirable activities • licensing agreements • standards for chemical selection and use • appropriate property rights for sustainable forest use

Source: Adapted from Pretty and Howes, IIED, 1993

Farmers and local communities increasingly need to be involved in forest investment and management. This reflects the reality that often forests are no longer solely in large blocks set aside and controlled by governments alone, but are located in many different areas, having been set aside or planted by different groups for different reasons. It is also in part because some forest communities, especially those with forest cultures that have evolved over centuries, are potentially good managers of forests for multiple purposes. Improvements will frequently entail ensuring appropriate rights to the forest, assisting with forest resource planning and management, and other means to strengthen farmer and community management. This will also help to integrate forests with surrounding agricultural and urban land uses.

The *private sector* can be in a position to better command the financial and technical resources required to protect, manage and harvest forest resources on a sustainable basis. By reducing commercial risk and requiring long-term commitment, certification may be one of the productive incentives needed for the private sector to invest in forests over the long term, as social and environmental assets.

Build capacities to meet current and changing forest needs

Improvements will be required in the way that government forestry departments are managed; and in the systems that the government uses to accomplish the tasks illustrated in Figure 4.1.

The more active involvement of additional stakeholder groups in planning forest activities, together with the devolution and privatization of certain functions which have traditionally been governmental, tend to be turning the role of government more towards that of facilitator, regulator and arbitrator. This shift in emphasis will require new and improved management, monitoring, participation and communication skills. It will be important to build, in parallel, the new capacities required to undertake agreed NGO and private sector roles. New kinds of forester need to be trained within and outside government: professionals who can deal better with the multiple dimensions of forests, who can handle the different stakeholder interests, and who can mediate the process of developing consensus.

Improve forest information, monitoring, valuation and research

Information collection and analysis needs to be increasingly well coordinated. This is so that appropriate sets of results can be made available to stakeholders and decision-makers, in a way which improves the policy cycle and forest management on the ground. At the national level, the main requirement is to bring information together on the supplies of forest assets, and the demands for forest goods and services, and to link this information to policy and planning cycles. This can be coordinated through central 'FRA' stocktaking for sustainable forest management and *valuation* systems. FRA is a means of keeping track of forest area, ownership, use, condition and management status (IIED/WCMC, 1994). Valuation systems should provide information on the economic, social and environmental values of forests, and be able to distinguish between changes in the capital value of forests and the income produced as goods and services. FRA and valuation can be linked. The government's annual report of forest activities could include an element of independent assessment as part of improving accountability.

National-level systems should be able to accommodate forest-level verifiable monitoring systems, to agreed standards. Mistrust between stakeholders at the forest level can, however, cause problems. Certification, by providing independent third party reports on forestry activities, may help to increase the confidence of stakeholder groups in each other.

Forestry research needs to diversify from its current production focus, to examine also the environmental, social and policy problems of forests. New research should relate more to policy needs; information systems; the forces acting on stakeholders, and their effects; the behaviour of forest ecosystems under different forms of management; and the ways in which different sectors should cooperate to achieve quality forestry.

Ensure country-level coordination of international forest initiatives

International initiatives concerning forestry, and related activities in agriculture, environment, natural resources and trade, are often uncoordinated. This can easily result in priorities being set outside the country. Countries therefore need to take the lead in better coordinating international interventions. This will entail:

- more active roles in forest issues by national agencies charged with intersectoral coordination – such as departments of planning;
- broader stakeholder involvement;
- acceptance by international agencies of the principle of country-level coordination; and
- stronger national involvement in developing and acceding to international programmes.

Improve the financial environment for forest conservation and management

Productive and sustainable forest operations need to have long-term financial viability, and to operate in a stable business climate. The costs required to meet the forest's long-term conservation requirements should be assured by fiscal independence of the government's forest authority. Governments should be capable of reinvesting fees obtained from forest utilization into maintenance of the forest resource. The tax base and subsidies required for forestry should be reviewed in light of the social, environmental and economic values of forests and of the long time scale of forestry operations.

International level

We have just described the importance of the national level in setting out policies to permit sustainable forest management. The question therefore arises: what are appropriate international roles? This is especially pertinent as international initiatives have often been ineffective, or have proposed excessive international control in the name of 'global forest conservation'. We suggest that there are three needs:

1 support to the all-important national processes;
2 dealing with legitimate global-level forest issues; and
3 agreements on the first two.

Some of the current initiatives, which go part-way to meeting these needs, are outlined in Chapter 9.

Certification in Context

International support for national processes

Principally this means transferring resources from rich to poor countries:

- finance for poor countries to cover the incremental costs of improving forest management; aid to cover long-term investment; and better coordination amongst countries providing this finance;
- **technical assistance** for capacity-strengthening and skills development;
- **sharing information, research and technology;**
- **harmonization** of data protocols and standards; and
- **improved trade measures** so that reforms in one country are not frustrated by fears of losing market shares to other countries.

Dealing with global forest issues

Principally this means mechanisms for handling those forest issues which have significant global implications:

- **setting principles and harmonizing standards for sustainable forestry to support trade in forest products:** international efforts are required to produce international standards in order to harmonize trade, but also to achieve consistency with environmental needs;
- **continued debate and dialogue** on global forest issues vis-à-vis national and local concerns; and
- **payments for 'global services':** this relatively new notion would entail supporting those activities which generate benefits beyond the borders of individual nations. Examples include the management of cross-border protected areas; carbon storage forests; areas of extremely high biodiversity; forest resources on desert fringes; and forests in regional watersheds.[2] The key issue is determining the extent to which costs exceed national benefits, ie how much payment is required to compensate for incremental global benefits.

Global agreements on the above

A number of global agreements already exist to support the above needs. These include CITES; the Biodiversity, Climate Change and Desertification Conventions; and the International Tropical Timber Agreement. As we discuss in Chapter 9, there is considerable debate on whether a Global Forests Convention is required. We conclude that it is only really essential for payments

---

2. Bilateral and private sector initiatives in this area already exist, such as that promoted by the FACE foundation in the Netherlands. FACE invests in forestry activities throughout the world, which have a net carbon sequestration effect, in order to mitigate the carbon emissions from electrical utilities in the Netherlands.

for 'global services'; and for encouraging or requiring involvement of all stake-holders in national-level forest activities – two relatively new areas upon which there is little consensus at present.

In this chapter we have looked at policies required to take countries through the transition to sustainable forest management. Many current policies were defined to serve narrow, static and simple ends, and by and large concentrated on government control. Policies for the transition to SFM will, however, be more dynamic and focus on groups other than government – notably the private sector and communities. With the involvement of many groups in SFM, the issue of standards becomes critical. This is addressed in the next chapter.

5. Standards

The issue of standards is one that has attracted considerable attention. It is bound up with the process of achieving consensus on a definition of sustainable forest management. As the previous chapter showed, this is an area where there is considerable difference of opinion; where emotions are high; and where results are often vague or ambiguous. Because of this, it has been difficult to produce standards for certification that follow proven practice and methodology developed for other processes and products. The result is often a muddle between general standards, which are external to the LFMU, and detailed standards which ought to be specific and internal to the LFMU.

Certification requires clear standards at the appropriate level, which are able to assist the practice of certification being precise, unambiguous and repeatable in its assessment.

In this chapter we clarify the terminology which is used in current standard-setting initiatives; and we show how a standards hierarchy should be correctly developed, so that even at the highest level of abstraction – usually at the international level – standards can be meaningfully interpreted for application at the site level. We show how principles already well-established under ISO could be used to help standard setting initiatives; and we show how current initiatives fit in with our proposed hierarchy.

External and internal standards

The term standard is used generically to describe a

> ... *quality or measure serving as a basis or example or principle to which others conform or should conform, or by which the accuracy or quality of others is judged*

as defined by the Oxford English Dictionary. Or

Standards are documented agreements containing technical specifications or other precise criteria to be used consistently as rules, guidelines, or definitions of characteristics, to ensure that materials, products, processes and services are fit for their purpose

as defined by ISO.

Within these all-embracing definitions there are different types of standard which can be categorized as being either external or internal to the LFMU. *External standards* are those set by third party independent bodies. These are usually international or national bodies; and all stakeholders usually participate in the standard setting process. *Internal standards* are those developed by LFMUs to describe the level of performance which their forestry activities must reach. Internal standards are interpreted from external standards and are specific to the LFMU. A vital part of the certification process is an assessment of this interpretation by the certification body. Figure 5.1 illustrates how this hierarchy looks with regard to certification.

Various attempts have been made to produce guidelines for forestry activities or broad statements that define sustainable forest management. These require considerable interpretation, and may provide insufficient guidance to certification bodies and forest managers for use at a field level. For external standards, it is usually necessary to develop principles and criteria. *Principles* should define the standard's scope. *Criteria* should set out the key elements or dimensions that define and clarify the principles. Both the principles and criteria – together with institutionalized checks and balances within the certification programme – should enable accurate forest-level interpretation by the LFMU and then the certification body.

Certification in Context

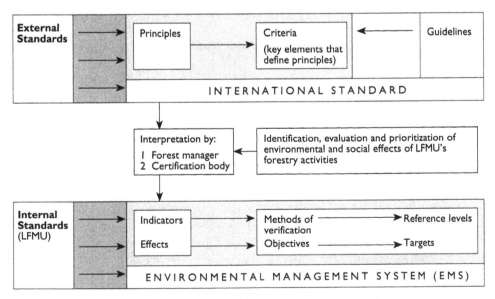

Figure 5.1 *The standards debate: terminology and hierarchy*

Certification in Context

For forestry, ideally a single external standard is developed at an international level for direct application by the LFMU. This, for example, is the practice followed with ISO standards. ISO standards are internationally recognized and directly applied by individual operations. The ISO standard for quality management systems – ISO 9000, details of which are given in Chapter 7 – is an example of a successful external standard which has international recognition and can be directly applied by local management.

As one of the objectives of certification is to improve the management of forests worldwide, and because wood and paper products are important in world trade, it is inevitable that external standards for certification will have to take international considerations into account. In particular, the standards that are set need to be compatible with free trade. However, for reasons of national sovereignty – that governments have traditionally been responsible for setting forest management standards and that the existing mechanisms for standard setting are often nationally based – the current tendency is to set standards nationally. Yet it is now considered by many that government involvement should not be exclusive and national standards should be set by a wider stakeholder group.[3] Conflicts of interest are perceived by some stakeholders when governments own and manage forests as well as regulate forestry activities. Where national standard-setting initiatives do proceed, it is important that mechanisms are examined to permit an eventual international harmonization of such initiatives.

In principle, a voluntary certification programme is open to all applicants. In practice, the development of inappropriate standards may discriminate against some applicants. The most common error in standard setting is to develop external standards that contain too much detail. Such standards are inflexible. Consequently, they are difficult to interpret and use in different forest situations; and they often discriminate against forests that were not in the minds of the standard setters when they were developed.

Detailed and inflexible external standards also tend to discriminate against poorer and relatively unsophisticated LFMUs – such as community-based forestry operations or farm woodlands. One can envisage a situation where a particular market that demands certified wood and paper products is soon supplied by wealthier and better-organized LFMUs which are able to respond relatively quickly. Poorer and less-organized LFMUs, which may have been traditional suppliers to this market, then have to withdraw because of their inability to respond to the new product specifications. Such a situation could further reduce the ability of such LFMUs to invest in quality forestry. These are also, probably, the forest areas where improvement is required most urgently.

In order to be compatible with free trade, external standards must allow flexibility to take account of different forest types, practices and objectives. External standards should not be overtly specific, which would effectively exclude imports of like products from being eligible. In order to ensure that flexibility in stan-

3 This is consistent with Agenda 21, the Forest Principles and other outputs of UNCED [1992] – see Chapter 9.

dards does not result in reduced performance by the LFMU to unacceptable levels, it is equally important that certification bodies use internationally-established norms for their operation; and that internationally–acceptable accreditation procedures are observed. Preferably these should be procedures already defined by ISO and used by recognized national competent bodies experienced in the regulation of certification programmes. These two issues are covered in Chapter 7.

There is, of course, a fine dividing line between trade discrimination and the competition between LFMUs which is required as part of certification. LFMUs need to recognize that certification can result in competitive advantage. A certification programme needs competition between LFMUs in order for certification to work as a market-based incentive. There has to be a balance between international acceptability, international fairness, site-specific implementation and an acceptable level of performance.

External standards which have been set at the international level can be complemented by various national activities. For example, accreditation of the certification body can be undertaken by national competent bodies. Accreditation of certification bodies to assess compliance with ISO standards is done in this way. In the UK, the NACCB (National Accreditation Council for Certification Bodies) is responsible for accrediting UK-based certification bodies wishing to issue ISO 9000 certificates. The NACCB accreditation can be used internationally in markets where it has credibility.

International principles and criteria can be both reinforced and complemented by appropriate national forest legislation; and by codes of forestry practice for specific forest areas. Such developments at a national level can play an important role in assisting effective site interpretation of external standards by the certification body and the LFMU.

External standards must be set with a clear goal in mind. For example, 'to improve the quality of forest management to achieve sustainability'. The standards would then relate to the key economic, social, environmental and management variables governing such management. The standards would also take account of *actual* best industry practices - rather than only an ideal - and required sector policy and institutional conditions.

Standard-setters must recognize that:

1   full information on critical variables probably does not yet exist;
2   standards and the performance of forest operations will improve over time; and
3   standards need to be simple and helpful, not complex and a barrier to progress.

There is more agreement on what is *unsustainable*, and on the basic elements of *quality* forestry, making it easier to agree standards on such a basis. Standards for *sustainable* forest management will evolve as experience with and confidence in certification progresses. As in other spheres, standard-setting should be evolutionary.

Internal interpretation of external standards set in this way should reflect a practical progression from:

1 cutting out unsustainable practice; to
2 introducing quality management that will form a basis for sustainable practice; to
3 identifying and achieving sustainable forest management itself.

In standard-setting it is vitally important to separate and clearly identify external from internal standards. In general, standards which refer to individual forestry activities are site-specific and should be developed internally. This includes standards which describe detailed road and bridge construction; or refer to minimum clear-cut sizes; or minimum girth limits for residual trees. External standards should set out general principles of quality forestry; including activities and documentary procedures which would be expected in a quality LFMU.

An exception to this distinction is *threshold issues*. Threshold issues are those which are considered important enough to be included in external standards, although they are site specific. An example would be the exclusion from use of certain types of chemical. The FSC, for example, specifically excludes the use of World Health Organisation Type 1A and 1B and chlorinated hydrocarbon pesticides whatever the circumstances. Apart from threshold issues, external standards should avoid reference to site specific activities.

External standards have to be acceptable to stakeholder groups. Participation is required to ensure that the perspectives and needs of different groups are included in setting external standards. An important consideration in so doing are the costs of implementing particular standards. In general, demanding standards are costly to meet; and detailed standards are costly to assess. Pitching the initial level required by external standards at just below current industry best practice ensures that they can be attained. Reducing the amount of detail in external standards will also reduce assessment costs, but should be accompanied by a strengthening of accreditation procedures.

Effective site interpretation of external standards is key to making certification schemes function properly. A first step is for the LFMU to declare its compliance with external standards in a *forest environmental policy*, a *sustainable forest charter* or a *code of practice*. An example of this is given in Box 5.1 which shows the 'green charter' of Demerara Timbers Limited in Guyana. Whatever the choice, this document must be endorsed and understood by everyone in the LFMU.

Interpretation of external standards in order to develop internal standards specific to the LFMU then requires accurate identification and prioritization of the environmental and social impacts of the LFMU's activities. This is usually achieved by undertaking an environmental impact assessment (EIA). The level of detail and coverage of an EIA should be appropriate to the complexity of the LFMU being examined. The EIA determines what are called *environmental effects* – using ISO terminology – but which have also been called indicators by others.

Box 5.1 *Demerara Timbers Limited – The Green Charter*

1   Each year, DTL will aim to harvest an area of approximately 30 000 acres at an average rate of 8m^3 per acre. This will give an annual yield of approximately 240 000 m^3. The 30 000 acres harvested will then be left for 20 years before further harvesting takes place. This extraction rate is the lowest figure proposed by expert forestry advisors.

They base their proposal on scientific research which has shown that natural growth in an untouched forest is about 1m^3 per 2.5 acres per year. The research also shows that if careful harvesting is carried out, this increases growth because the forest is opened up.

2   DTL will fund independent scientific research to confirm that these findings also apply in its forest area.

3   When DTL plans the harvesting, it will ensure that trees are never harvested in such a way that large openings occur in the canopy.

4   DTL will plan harvesting to ensure minimum land use for roads.

5   When roads are needed, DTL will ensure that they are aligned to the natural contours of the terrain to avoid soil erosion.

6   DTL recognizes that it will not be alone in its forests. DTL will specifically propose to the Government of Guyana: (a) that large-scale farming will be banned for the life of DTL's forest lease; (b) that hunting be reduced to a minimum; (c) that any mineral licences granted should be accompanied by strict environmental controls.

7   DTL will, over the life of its lease, make a special effort to develop and promote species other than those which are presently considered commercial. The aim will be to reduce the harvesting of the most popular species as new species become accepted in the market place. DTL's marketing programme will cover 18 species from the outset.

8   DTL will aim to increase its product range to reduce waste.

9   DTL will work towards a waste management programme to ensure that waste is being converted into charcoal, briquettes or energy.

10 DTL will seek to establish in Guyana traditional downstream activities, such as furniture production, so that more value is added to Guyana's forest products, more jobs are created and DTL's earnings are enhanced while the company makes its optimum contribution to the nation's economy.

Source: Demerara Timber Limited, Guyana

Certification in Context

The LFMU's environmental effects should be identified, evaluated and listed. From this information, the LFMU should establish its objectives and targets – again using ISO terminology. Others have called these *means of verification* and *reference levels*. The framework for establishing internal standards, which is described above, follows that used in implementing an environmental management system. Chapter 7 provides a more detailed discussion on environmental management systems, as these are central to the ability of LFMUs in achieving certification.

Objectives are the broad goals that the LFMU sets itself to achieve, and which are quantified wherever possible. They should reflect the scope of the external standards and the identified environmental effects. *Targets* are detailed performance requirements, quantified wherever practicable, and applicable to the environmental effects of the LFMU. Targets are set in order to ensure that objectives are met.

Targets are, in effect, detailed site-specific standards, the application of which should result in immediately acceptable levels of social and environmental impact resulting from the forest organization's activities. Regular internal monitoring to measure the organization's performance against the targets allows for priorities and objectives to be re-examined; and provides the basis for continuous improvement.

For example, an objective might be to leave sufficient standing stock after harvesting to ensure regeneration. The target might be to specify that all trees of less than 30 cm DBH (diameter at breast height) be left standing. Where a particular target is not met, the certification body agrees a time limit with the LFMU for this to happen. Fulfilment of the agreement to the satisfaction of the certification body may be conditional for award of the certificate where importance of the target is significant; or conditional to maintaining the certificate where importance of the target is less significant.

Internal standards should provide the targets for making the transition, to an ideal of sustainable forest management from the current situation. Application of the standards should result in an immediately acceptable level of social and environmental impact directly associated with the forestry activities of the LFMU and external to it. We have defined this as quality forestry.

Derivation of internal standards should mean that they:

- ensure that forest practices attain minimum acceptable levels of performance;
- provide a benchmark against which forest management performance can be measured and further improved; and
- reconcile the mosaic of site-specific differences within each forest which may require different approaches to forest practice.

A number of problems currently exist in defining internal standards; these are discussed below.

Lack of information on quality forestry

A problem associated with the development of internal standards is a lack of scientific understanding as to what is, and what is not, quality forestry in a given area. Especially in the tropics, basic knowledge such as the growth rates of particular harvested species, the reproductive strategies of lesser-known species, the existence of rare or endangered species, and the identities and requirements of 'keystone' species that are crucial to ecosystem functioning, is often not known for a particular area. This knowledge is, however, required to identify correctly the nature and extent of an LFMU's environmental effects. Such information is also required to make the step from overall objectives set by the LFMU to the specific targets required for quality forestry to be achieved. This problem is an important reason why certification has to be dynamic and allow for improvement over time.

Standards, and tests for the assessment of those standards, are intimately linked. Assessment procedures should also be based as far as possible on verifiable fact. However, there is often a lack of sound information on the relationship between particular standards and well-defined scientifically supportable tests for these standards. Very often this relates to the lack of basic data from which to develop the standards themselves. For example, without adequate knowledge of the growth rates of major timber species in a particular forest, it is impossible to provide accurate and well-defined limits of exploitation, and to develop objective, repeatable tests to assess breaches of those limits.

Even with current knowledge, however, it is possible to develop basic internal standards which if implemented would lead to a great improvement in the quality of management of many forests; and assessment procedures to test compliance with such standards. Although it is difficult to define the scientific limits of sustainable forest management, it is a straightforward task to identify inadequate management practices that can be improved.

Making trade-offs between economic, social and environmental objectives

Managers of LFMUs and certification bodies are required to come to a judgement regarding certification, knowing that certain environmental and social objectives may at times have to be temporarily compromised by the financial demands of the LFMU; or by the need to adjust to major events such as fires, insect attack, disease and storms, which severely disrupt the diversity and age distribution of the growing stock.

To cope with such situations, it is important that both the LFMU and the certification body incorporate appropriate procedures into planning and management systems. An assessment of the adequacy of management systems to cope with such situations should be an integral part of the certification process.

To summarize – good standards should satisfy well-defined and realistic boundary conditions; and be consistent and harmonized between external and internal levels and between different forest types.

How standards are set by ISO

The procedures for standard-setting have been well-established by a number of national and international organizations – notably the International Organization for Standardization (ISO) – see Box 5.2.

Box 5.2 *International Organization for Standardization (ISO)*

ISO is a worldwide federation of national standards bodies from 90 countries established in 1947. It is an NGO with a mission to promote the development of standardization to improve the international exchange of goods and services, and to develop intellectual, scientific, technological and economic activity. Its work results in international agreements which are published as International Standards.

ISO derives from the Greek '*Isos*' meaning 'equal' and, as an acronym, is valid in each of the organization's three official languages – English, French and Russian.

Member bodies of ISO are the national bodies most representative of standardization in its country. Only one such body for each country is accepted for membership. The member bodies have four principal tasks:

1   informing potentially interested parties in their country of relevant international standardization opportunities and initiatives;
2   ensuring that a concerted view of the country's interests is presented during international negotiations leading to standards agreements;
3   ensuring that a secretariat is provided for those ISO technical committees and sub-committees in which the country has an interest; and
4   providing their country's share of financial support for the central operations of ISO, through payment of membership dues.

There are two other categories of member. A *correspondent member* is usually an organization in a developing country which does not yet have its own standards body. *Subscriber membership* is provided to countries with very small economies. These subscribers pay reduced membership fees, which nevertheless allow them to maintain contact with international standardization.

The scope of ISO is not limited to any particular fields. It covers all standardization fields except electrical and electronic engineering. The technical work of ISO is highly decentralized, carried out in a hierarchy of some 2700 technical committees, sub-committees and working groups.

Examples of ISO standards include:

• universal system of measurement known as SI;
• paper sizes – the original standard was published by DIN (*Deutsches Institut für Normung*) in 1922 and is now used worldwide as ISO 216;
• internationally standardized freight containers which enable all components of an international transport system – air and seaport facilities, railways, roads and packages – to interface efficiently;
• the diversity of screw threads for identical applications (which used to represent an important technical obstacle to trade and maintenance). A global solution is supplied in the ISO standards for ISO metric screw threads.

ISO standards are developed by technical committees which should include qualified representatives of industry, research institutes, government authorities, consumer bodies, NGOs and international organizations from all over the world.

The major responsibility for administering a specific standards committee is accepted by one of the national standards bodies that make up the ISO membership. The member body holding the secretariat of a standards committee normally appoints one or two persons to do the technical and administrative work. A committee chairman assists committee members in reaching consensus. Generally a consensus will mean that a particular solution to the problem at hand is the best possible one for international application at that time – since the membership tends to encompass those with knowledge of the necessary policy, technical and operational issues. Each member body interested in a subject has the right to be represented on a committee. International organizations, governmental and non-governmental, in liaison with ISO, also take part in the work.

The ISO Central Secretariat in Geneva ensures the flow of documentation in all directions; clarifies technical points with secretariats and chairmen; and ensures that the agreements approved by the technical committees are edited, printed, submitted as draft International Standards to ISO member bodies for voting, and published. Meetings of technical committees and sub-committees are convened by the Central Secretariat. Although the greater part of the ISO technical work is done by correspondence, there are, on average, a dozen ISO meetings taking place somewhere in the world every working day of the year.

ISO's standards are prepared according to the following principles:

- **Consensus:** through committees, working groups and document distribution; the views of all interests are taken into account (manufacturers, retailers and users, consumer groups, assessors, governments and professionals).
- **Industry-wide:** global solutions are sought.
- **Voluntary:** as international standardization is market-driven, it is based on voluntary involvement of all the interests in the marketplace.
- **Periodic revision:** technological evolution, new methods and materials and new societal requirements for quality and safety mean that ISO standards are reviewed at least every five years.

There are three main phases in the ISO standards development process.

1 The need for a standard is usually expressed by an industry sector which communicates this need to a national member body. The latter proposes the new work item to ISO. Once the need for an international standard has been recognized and formally agreed, the first phase involves definition of the technical scope of the future standard. This phase is usually carried out in working groups, which comprise technical experts from countries interested in the subject matter.
2 Once agreement has been reached on which technical aspects are to be covered in the standard, a second phase is entered during which countries negotiate the detailed specifications within the standard. This is the

consensus-building phase.

3　The final phase comprises the formal approval of the resulting draft international standard. The acceptance criteria stipulate approval by two-thirds of the ISO members that have participated actively in the standards development process, and approval by 75 per cent of all members that vote. Following this, the agreed text is published as an ISO international standard. This can be anything from a four-page pamphlet to a 1000-page tome.

Many standards require periodic revision. Several factors combine to render a standard out of date: technological evolution; new methods and materials; and new quality and safety requirements as a result of changing market and legislative demands. To take account of these factors, ISO has established the general rule that all ISO standards should be reviewed at intervals of not more than five years. On occasion, it is necessary to revise a standard earlier.

ISO standards can be divided into those which directly concern *products* and those which concern *processes*. Accreditation and certification practices differ with regard to assessing compliance with these two types of standard. In the UK, for example, there are different accreditation bodies – NAMAS (National Accreditation and Measurement Advisory Service) for certification of laboratories testing for product conformity; and NACCB for certification bodies assessing systems (process) compliance. Standards for processes also require greater levels of professional judgment on the part of the certification body, and empirical testing procedures are unlikely to be available.

As forestry is an activity rather than a product, certification has most in common with process standards. Examples of these include ISO 9000 for quality management systems; and ISO 14 000 – currently in draft form – for environmental management systems. Both these standards are primarily concerned with systems rather than performance or products. Indeed this has been one of their main criticisms. In assessing compliance with the standards, certification bodies have to use specifically-trained staff; and have detailed procedures of their own in place to ensure that assessment is repeatable and takes place within an adequate framework. Accreditation of certification bodies places emphasis on these points.

The important point to recognize is that, where assessment of compliance with standards requires significant professional judgement, accreditation rules for certification bodies need to be carefully thought out and rigorously applied. As the previous section has demonstrated, because of the site-specific nature of forestry activities, external standards will always require significant site interpretation, using professional judgement first by the LFMU and second by the certification body. For this reason, in forestry, the accreditation procedures for certification bodies form an important part of the certification process.

Existing initiatives in standard setting

In certification, considerable effort has been spent on defining what the standards of sustainable forest management should be. Solutions which can be agreed at a theoretical or philosophical level may have practical limitations. In addition, a relatively large amount of time has been spent on producing an 'ultimate' set of standards – but this work has tended to proceed in isolation from describing and regularising an institutional and regulatory *structure* for certification; and the development of effective assessment *procedures*. While most of the institutional and regulatory needs are already defined by ISO standards, the linkage with appropriately detailed standards and effective assessment procedures makes it important to consider these three aspects at the same time.

One effect of the emphasis on producing standards is that several versions have been developed by a wide range of organizations. At present, nearly every certification body and nascent certification programme defines their own standards. One result is that certificates from different certifying bodies for similar forest areas refer to different standards. However, as certification develops, the rigours of accreditation and practical experience will force a convergence of standards. This is already happening with the four up-and-running certification bodies (Chapter 11) all maintaining programmes that are compatible with the Principles and Criteria established by the *Forest Stewardship Council (FSC)*.

In the international arena, only the FSC has produced standards which are explicitly for certification (see Chapters 9 and 12 for details on the FSC). Indeed, the FSC was established to develop certification internationally. The FSC has consistently insisted on wide consultation during its formation and the development of its standards. For most of 1993, the interim board of the FSC organized an international consultative process which took place in ten countries. This process culminated in a founding meeting during October 1993; and ratification of the first set of standards in June 1994.

The FSC standards are designed for all types of forest: boreal, temperate and tropical as well as natural forest, plantations and conversion forest. They are expressed as Ten Principles and Criteria, the text of which for natural forest was agreed and ratified in June 1994. Principles and Criteria for plantations and conversion forest have yet to be finalized, though that for plantations is awaiting board ratification. The FSC also expects to commence formal accreditation of certification bodies later in 1995.

The FSC Principles and Criteria form a broad umbrella under which specific standards can be formulated by certification bodies and LFMUs for different forest types in different parts of the world; and by which these forests can be consistently evaluated.

Despite the effort at consultation, which is still ongoing, some groups consider the standards to be too demanding on environmental and social factors, and yet weak on the need for economic viability. Many economic groups feel alienated by the consultation process and unrepresented in decisions. In particular, they are concerned by the lack of a realistic and explicit provision for improvement over time. However, the FSC does aim to ensure that initial local standards are not so high that they define only an 'elite' few forests, but not so low that standards are meaningless. Periodic revision and an insistence to assess change over time (FSC Principle 8) will ensure that quality is continually raised and allows for some initial flexibility.

Although still in draft form, the *Canadian Standards Association (CSA)* set of standards for Sustainable Forest Management (SFM) is an example of a model which is designed to fit in with ISO practices and allows for improvement over time within a defined and assessed environmental management system (EMS). The importance of EMS and its relevance to certification is elaborated in Chapter 7.

Two draft standards have been developed by the CSA:

1 Z808 – Sustainable Forest Management; and
2 Z809 – Sustainable Forest Management Systems.

The CSA also intends to develop a document 'Guidelines for Sustainable Forest Management Auditing' (Z809.1). The CSA standards aim to incorporate the merits of a systems approach to forestry, as defined by an EMS, and which sets an acceptable level of performance through specified principles and criteria of SFM. To comply with the CSA standards the LFMU must demonstrate firstly, that its management system is in conformity with the standard; and, secondly, that the level of environmental and social performance achieved by the management system complies with the specified principles and criteria. The CSA initiative is also designed so that the defined principles and criteria could be made compatible with an acceptable internationally-defined standard. Initiatives cited include: The Helsinki Process, the Montreal Process, the International Tropical Timber Organisation (ITTO) and the United Nations Commission on Sustainable Development. The current draft of the standard bases its principles and criteria on the essential environmental, social and economic values agreed by the Canadian Council of Forest Ministers (CCFM) at the beginning of 1995.

Environmental groups, who generally support the FSC, interpret the absence of any recognition of the FSC in the CSA standards as a sign of hostility. They are also concerned that the essential values agreed to by the CCFM allow for too loose an interpretation and, therefore, provide an inadequate assurance that the standard's prescribed management system will reach the required level of performance – ie that the management system approach allows for too little improvement over too *much* time.

In practice, there appears to be little difference in the scope covered by the FSC and the CSA (see Chapter 9). It remains to be seen whether the accreditation rules for certification bodies will provide greater assurance that the essential values agreed to by the CCFM will be interpreted adequately. In the meantime, the CSA's contribution of using an EMS framework to deal with the site-specific nature of forestry operations should be acknowledged.

There are other international initiatives which have produced forestry standards. Perhaps the most comprehensive of these are the *ITTO Forest Management Guidelines* for the management of natural forest and, separately, for the management of plantations. Criticism of the ITTO guidelines has focused on their exclusion of forests other than tropical; their lack of adequate social parameters; and that they were drawn up without sufficient consultation. However, the ITTO Guidelines recognize many of the wider aspects of forestry that ITTO member countries have been widely criticized for ignoring. They aim to provide an 'international reference standard' for 'all producer and consumer countries which are concerned with the efficient and sustainable development of the tropical forest resources and forest-based industries'. As such, they are a benchmark for the basic controls that producer governments must apply in order to achieve quality forestry in the tropics. Further development is being encouraged by ITTO to produce standards that are applicable to specific countries or forest types, and to develop ways of interpreting these standards which can be applied by the forest manager. ITTO and certain member countries are currently involved in developing such nation-specific criteria.

Extremely detailed standards for tropical forests have also been developed by *Initiative Tropenwald (ITW)*, based in Berlin, Germany. During 1995, CIFOR will test the effectiveness of ITW standards alongside those of Smart Wood, Woodmark, LEI (Indonesia) and the Helsinki process in a variety of forest environments. The first test, undertaken at Bovenden in Germany during January 1995 showed the field assessment of detailed external standards to be time-consuming and suggests that the approach taken by ITW will result in costly assessment. Other tests will take place in Indonesia, Côte D'Ivoire, Brazil, and possibly Cameroon.

Other international forestry standards have been developed independently by certification bodies such as the Soil Association in the UK and the Rainforest Alliance in the USA. In common with FSC's Principles and Criteria and ITTO's Guidelines, there tends to be a presumption that a closeness to natural, unmanaged forest is indicative of sustainability –polycyclical, highly diverse systems are often chosen as the standard. This may prejudice against some monocyclical, low-diversity systems which otherwise achieve considerable social, environmental and economic benefits.

This drawback has been recognizd in the case of plantations, for which separate guidelines have been drawn up by ITTO and a separate principle is now being finalized by the FSC. The Smart Wood certification programme operated by the Rainforest Alliance appears to recognize implicitly such differences by operating a two-tier system – recognizing first good management and then sustainable management. No source, as yet, has been certified through the Smart Wood programme as sustainable.

Certification in Context

Remaining dilemmas in standard-setting

- **Getting external standards adopted in general procedures and practice.** There is sometimes a misfit between *existing national regulations* for forestry, which have developed organically over the years, and *standards currently being developed*. Because of this, the next few years may see a rapid evolution of national regulations and procedures to accommodate currently acceptable standards. How certification can contribute and work within this process has yet to be addressed. However, it may transpire that this development is one of the greater indirect benefits from certification.
- **Dealing with conversion forests.** Current standards tend to look within the forest, and do not address the balance of forest in land use. Most notably, standards for conversion forests do not yet exist. It should be legitimate for many countries to fell certain forests, to put the land under alternative uses (tree plantations, agriculture, etc) provided such decisions were the result of rational land use planning which took account of stakeholder needs for sustainable development. Existing standards emphasize the continuity of forest cover and are not equipped to handle the possibility of well-managed forest clearance. Hence products from conversion forests would not, at present, meet any existing standards and could not be certified.

The challenge of the eco-label

There is increasing interest in life cycle eco-labelling of many forest products. Such labels cover standards not only for forest management, but also for pulping, packaging, manufacture, transport, use and disposal. Examples include the European Commission regulations which establish criteria for the eco-labelling of toilet paper (Reg: 94/924/EC) and kitchen rolls (Reg: 94/925/EC).

Forest certification is often presented in a way which makes it synonymous with eco-labelling. This is confusing, as there is an important difference. Forest certification is more correctly defined as a *single issue* eco-label – wood and paper products are labelled according to whether the forests they originate from are well managed. Forest certification allows the use of an eco-label which only considers the production of particular raw materials – virgin fibre for paper products and timber for products using wood.

More usually, eco-labelling is *multiple issue* and adopts a cradle to grave analytical framework (life cycle analysis or LCA). The starting point for a life cycle analysis is the correct identification and prioritization of environmental effects throughout a product's 'life cycle'. Typically, a matrix is prepared which ranks the relative importance of environmental effects according to each phase of the product cycle. Criteria which set standards for each effect identified as significant are then developed. An acceptable identification of effects is one of the main challenges facing LCA for wood and paper products. A credible forest certification programme could be of assistance in facilitating this work.

A forest certification programme could fulfil part of the requirements for a multiple-issue eco-label, provided mutual recognition between the two programmes existed. For example, pulpwood originating from a forest certified under a Swiss forest certification programme could form part of the EU eco-label for toilet paper. Current eco-labelling initiatives should consider ways in which existing and credible forest certification programmes can be accepted. Such recognition would facilitate the development of eco-labels and reduce the cost of their application.

6. What Certification Achieves

At present, many forests are probably not certifiable without sufficient time being allowed within the certification process to make the required improvements. Legislative and policy reform may also be necessary to give forest managers confidence in reaping the benefits from increased investment in forestry activities with long-term returns.

Within this context, the critical question is: *how can certification provide an adequate incentive for a large-scale transition towards sustainable forest management?* Is certification destined to satisfy the needs of only a few 'elite' forests able to satisfy a niche 'green market', where price premiums for sustainably-produced products may exist?

In beginning to answer these questions, against the debate of the previous chapters, we look first at the effectiveness of certification as a market-based instrument; and then the extent to which certification can address the forest problems identified in Chapter 2. In appraising its effect on forest problems we also consider where the benefits of certification can work directly; or where its main benefits may be indirect.

Certification: a market-based instrument

Certification is an economic market-based instrument which aims to raise awareness and provide incentives for both producers and consumers towards a more responsible use of forests. It is an especially appropriate choice of instrument where:

- there is a strong willingness amongst consumers to pay the extra costs associated with the instrument. This usually takes the form of higher prices, which may be achieved where additional environmental aspects are recognized as enhancing product quality; or

- any associated increase in cost associated with the instrument is offset against other commercial gains. In certification these could include:
  a) medium-term gains in efficiency and productivity
  b) protection of market share and increased marketing opportunities through product differentiation
  c) reduction of environmental risk, resulting in better access to financial markets for loans, rights issues, insurance, etc
  d) better stock control
  e) improved image in 'green' conscious markets and with employees.

Such instruments can require supplementing by regulations or other controls to ensure that the products they aim to promote are not at a market disadvantage compared to less environmentally sensitive substitutes. However, for organizations involved in certification, the benefits from sales of certified product may not come from specific product benefits – such as higher prices. The real benefits may come from an improved business profile. In certain markets 'green' is associated with the 'attitude' of the producer more than the content of the product. Especially in North America and Europe, growing numbers of consumers and employees seem to be attracted to and interested in 'green' issues. They will identify more easily with those organizations which convincingly demonstrate concern for 'green' issues; and which sell and promote 'green' products. This can have benefits in terms of improved *overall* commercial performance.

The benefits from market-based instruments are low where people do not demand the product, where they cannot pay for the product, or where markets are undeveloped. In this respect, experience with marketing a range of 'green' products permits some generalities to be made which are of relevance to certification:

- Consumers will not pay more for certified products *where product quality and performance is lower than that of alternative products.* It is unlikely that many consumers will even pay the same price for certified products where this is the case.
- To command price parity with alternative products, it is likely that certified products must equal the quality and performance of such products. *Where certified products can equal or exceed the quality and performance of alternative products* then their identification as a certified product may encourage increased sales.
- *Even where the quality and performance of certified products equals or exceeds that of alternative products*, there is no assurance that consumers will pay increased prices, except in a few 'niche market' instances.
- *Green labelling can differentiate and improve market share for a product.* Where the satisfaction of environmental issues is widely considered to form part of a product's quality specification, green labelling can be perceived as improving the quality of that product. Improving the quality of a product and selling it for the same price can help in maintaining or even increasing

---

Box 6.1 *The 'green market' pull*

The 'green consumer' emerged during the 1980s, almost entirely in richer countries, and since then has exerted pressure in a variety of sectors. There have been some notable successes such as the switch from using CFCs and leaded petrol.

Many products have undergone the scrutiny of environmental groups [NGOs] and the idea of providing the consumer with choice based on environmental criteria is not new. Survey work conducted by the UK Consumers Association, on environmental labelling in general, revealed that 19 per cent of the adults in the survey listed environmental concerns as an important – but not the main – consideration when making purchasing decisions. The other 81 per cent said they were prepared to pay *more* for goods which are environmentally less damaging.

The future of the world's tropical rainforests has been for some years now a major public concern in richer, often temperate, countries. NGOs have provided information to the public and the media on the potential seriousness of deforestation. Rightly or wrongly, trade and industry are often depicted as key agents in tropical forest destruction; and the concerned consumer is encouraged to use his/her power to address the issue of deforestation. Children's books explain the causes of rainforest demise and the children are 'asked to tell their parents not to buy furniture made of rainforest trees'.

Recently the debate has also turned its attention to temperate and boreal forests, notably the felling of old-growth forests such as at Clayoquot Sound in British Columbia, Canada.

In addition, NGO groups have put direct pressure upon selected traders, particularly in the UK DIY [do it yourself] chains and other retailers who have high visibility among customers. NGOs have used harassing tactics including barricades and demonstrations. In response, retailers are endeavouring to meet perceived customer demands; or to safeguard their market share and their image. The response often appears more a reaction to avoiding environmental risk than searching for increased sales through the promotion of 'green' goods. 'Any successful retailer fulfils one basic principle: that is we meet and exceed our customers' demands and expectations ... environmental issues are becoming more and more of a concern to our customers.' (*Alan Knight, B&Q*) It is important to try to analyse the situation more closely in order to assess the true extent of these 'green trends'.

The current global volume of certified timber (about 1.5 million m^3 per year) is marginal in view of the total size of the market. According to certifiers, there is more demand than supply in their primary markets. But how easy is it to reliably quantify the demand for certified wood and paper products? According to a recent survey, 89 per cent of UK timber traders reported that a very few, or none, of their customers enquire about the source of timber products [*Timber Trade Journal*, 1994]. How does this square with the fact that retailers appear to be increasingly considering environmental factors?

Caution is needed when assessing the situation. NGOs are very good at identifying and targeting key groups which are affected very directly by public opinion. These include local political bodies, retailers with a high public profile, stock exchange listed companies, etc.

Market demand for green products is not even, and there is a great difference between consumer awareness and the willingness to act on this awareness. This also varies from country to country. Countries outside Europe and the US are experiencing very little environmental pressure. Within Europe, the UK, Germany and the Netherlands demonstrate a high level of awareness compared to France, Italy and Portugal, where environmental concerns are low – although increasing.

However, some European countries, regardless of national sentiment, are being forced to consider the attitudes of consumers elsewhere. In Italy, large volumes and values of tropical timber are used to make furniture which is exported to other European countries. The Italian timber trade is therefore highly exposed to public attitudes in export markets.

Studies have shown that there is no convincing evidence that a price premium exists for certified timber. Customers may say they are willing to pay more for certified timber, but will they *actually do it?* Important factors are: levels of disposable income; the stage of a particular country in the business cycle; and geography. It is more likely that consumer environmental awareness will have a greater impact on the market share of tropical wood compared with wood from temperate areas; and of wood in general against substitute products.

Fashion and the level of economic activity appear to be more important in determining market access than purchasing decisions based on environmental criteria. A slowdown in the construction industry is generally considered to have reduced recent tropical wood imports into France. The current fashion for lighter-colour furniture has contributed to an increased demand for temperate hardwoods compared to their generally darker tropical counterparts.

A problem faced by all 'environmentally friendly' products has been highlighted by the UK Consumers Association. To what extent do the public understand the labels and the claims? A survey looking at the customer's perception of environmental labels in general revealed that most people (55 per cent) thought the label implied some sort of official approval. Among those who knew that labelled goods do not require official approval, 83 per cent felt they should require it. Overall the preference was for Government departments to take responsibility for approving environmental labels: nearly 60 per cent of consumers favoured Government-supervised labelling: no other organization came close. Sixteen per cent would allow manufacturers to issue the labels themselves and the EU was named by only 1 per cent.

The facts and figures

- In 1989 **B&Q** (a DIY chain in the UK) surveyed their customers and found 76 per cent were concerned about rainforest condition and 15 per cent would refuse to buy tropical timber.
- A 1991 MORI poll commissioned by **WWF** revealed that 48 per cent of the UK population regarded themselves as concerned about rainforest destruction and that 15 per cent try to avoid buying tropical hardwoods.
- In the **US** Winterhalter and Cassens (1993) made a study which revealed that, in a sample of 12 000 consumers having annual incomes of more than $50 000, 68 per cent of consumers were 'willing to pay more for furniture whose construction material originated from sustainably managed North American forest'. Knowing the origin of the material was the fourth most important selection criteria among the six proposed.

Certification in Context

- In 1994, a survey undertaken by **Mintel International** showed 60 per cent of UK shoppers prepared to adapt their purchasing habits according to their environmental principles; and 40 per cent of the total make a positive effort to buy environmentally friendly products. Only 10 per cent of shoppers admitted to having no concern for the environment. However, more than half of the respondents found green claims confusing.
- In the **UK**, many local councils operate procurement policies with respect to tropical timber purchasing, as does the Ministry of Defence. In the **US**, the Minneapolis City legislature is considering a ban on the use of all tropical timber within the city limits.
- In the **Netherlands** a survey of the 50 largest towns and cities conducted by the Dutch Friends of the Earth revealed that: 6 local councils have to a certain extent prohibited the use of tropical hardwoods; 35 local councils refrain from using tropical hardwoods in projects commissioned by themselves; and 26 local councils have a 'kind of policy regarding tropical forests'.
- In **Germany** there are bans against the use of tropical timber in many large cities such as Cologne, Frankfurt, Hamburg, Bremen and Munich and also, according to the environmental group Rettet den Regenwald, an unknown but steadily growing number of smaller communities.
- A recent publication by the **UK Institute of Directors**, which has 48 000 individual members, said that businesses should not buy products that 'destroy tropical rainforests'.
- Presently, because little certified timber is available in the **Netherlands**, customers will, where appropriate, use recycled plastics – paying up to four times as much for this option. It is believed that a price premium of 5–10 per cent on certified timber will be acceptable.
- In **Switzerland**, it is generally assumed that the reduction in tropical imports by nearly 50 per cent since the mid eighties is the result of environmental pressure.

market share in mature markets. Where consumers are wary, for environmental reasons, of purchasing wood products which are not identified as certified; then green labelling of wood and paper products can assist in maintaining market share against substitutes.

There is a plethora of conflicting evidence on the market for certified wood and paper products, and on the extent to which consumers are prepared to pay a 'green' premium. Box 6.1 discusses specific aspects of this market and presents some of the results obtained from surveys related to wood products. Experience from these and the marketing of other 'green' products suggest that, in all likelihood, most buyers are *not* prepared to pay a price premium for certified wood and paper products. However, despite this, retailers and local authorities seem keen to insist upon supplies of certified wood for other commercial or political reasons.

The main commercial effect of certification instead probably revolves around environmental risk. Because of mainly public and environmental NGO pressures, companies exporting to and selling wood and paper products within certain North American and European markets are increasingly identified as agents in forest degradation. Rightly or wrongly, and particularly where the wood and paper products concerned are of tropical origin, these companies are perceived as carrying high levels of environmental risk. As concern for environmental issues continues to develop, perceptions of environmental risk are likely to play an increasingly important role in commercial decisions.

In the future – if not already – failure to reduce this environmental risk is likely to result in increased cost. This will affect the commercial returns of those companies identified as having both direct and indirect impacts on forests. Increased cost and reduced commercial return are likely to be the result of several combined factors. These may include:

- poor environmental image;
- difficulty in maintaining market share and securing new markets;
- difficulties with third parties in securing permissions – such as for new buildings;
- low staff morale, increased staff turnover and loss of good staff to competitors; and
- increased insurance and financing costs.

Where they exist, the negative commercial effects of increased environmental risk are likely to develop gradually over time. For many companies, the real nature of the problem may not become clear until some damage has already occurred. Certification and its promotion provides a mechanism for companies to reduce environmental risk and the negative commercial effects that high environmental risk increasingly involves.

Seeking to mitigate the negative effects of environmental risk through certification is the mirror image of seeking to improve commercial and competitive advantage through certification. The perspective is different but the process is the same. Thus the effect of certification can be viewed as either a positive or negative incentive.

As a *positive incentive*, certification has to provide the company involved with a commercial advantage over its competitors. This can take the form of preferential access to new customers or increased market share, or better prices through direct sales or niche marketing.

As a *negative incentive*, certification ensures that the costs of not participating are greater than those of participating. This usually means that, although participation in principle is voluntary, the company feels compelled to comply for fear of losing market share and access.

The main protagonists have yet to make a convincing case for certification as a suitable market-based instrument in its own right. The principal rationale is based on the historic failure of alternative options. The principal assumptions are that regulatory approaches, often government-inspired, have failed; that the 'green' consumer now exercises significant purchasing decisions in favour of 'green' goods; and that the enlightened self-interest of the corporate sector in the face of declining resources and increasing client power have created the conditions for market-based certification programmes to succeed.

This has been expressed by WWF as follows:

> *In the absence of a legally-binding international forest agreement, we believe that industry has a responsibility to ensure that the timber it uses comes from well-managed forests – for two reasons. Firstly, any company which ignores the sustainability of supply of its raw material base is very vulnerable, especially if, as with forests, the resource is being over-exploited. Secondly, companies have a responsibility to carry out their business in a way which reduces their impact on the environment. Shareholders and customers are increasingly demanding this.*

> *WWF, January 1994*

The market aspects, versus regulatory aspects, of certification highlight a dilemma: are law and regulation best suited for upholding society's wishes; or are these mechanisms, stuck in a bureaucratic inertia of their own, too far behind society's desires? If so, can society's desires now be better expressed through the marketplace and by the voluntary efforts of industry? Is enlightened self-interest enough to ensure long-term survival of forests?

Market-based, voluntary approaches do appear to have certain advantages:

- They can yield *cost savings* over regulatory mechanisms, by allowing forest managers to determine the most appropriate and cost effective ways of meeting given external standards.
- They offer an *ongoing incentive* to continuously improve, and in so doing encourage the development of new techniques and technologies.
- They are *flexible*, allowing forest managers to work within an overall constraint and do not require the drafting of new legislation and regulations by government.
- They are relatively *efficient* in terms of resource pricing, and thus can promote conservation and long term sustainable resource use.

However, there are some operational dilemmas:

- There is much *more experience of using regulation* than there is of economic instruments to achieve environmental and social goals. Economic instruments have been much talked about, but little implemented.
- Economic instruments are unlikely to work on their own in *low-income or non-market situations*. This is where people are too poor to pay; and where markets are undeveloped, facing many uncertainties about supply and demand and macroeconomic instability.
- In the past, the market can hardly be said to have been the friend of the forest. Rather, the *commoditization* of just one or two forest goods, wood and

land, *has led to widespread forest loss and degradation*. Some observers even feel that this has been the main cause of forest loss and degradation.

Hence regulatory and institutional provisions are likely to be needed as a supplement to market-based initiatives such as certification to correct these market failures.

Certification is expected to work through the market. Forest managers who produce goods and services that a) do *not* enter the market or b) enter undeveloped markets or markets which do not discriminate in favour of products from well-managed sources, are unlikely to see any immediate value in certification. Certification may be most successful as a market-based instrument in well-developed markets of high-income countries, but not for many low-income countries. In markets where certification can succeed, early entrants may gain a competitive advantage.

Where wood and paper products are traded, it also follows that certification could succeed for the sale of such products from low-income to high-income countries, but not, under current conditions, for sales between low-income countries. Low-income countries may need assistance in order for their timber to compete with timber from high-income countries on an equal footing in high-income country markets. Many low-income timber-exporting countries are predominantly tropical and may not have the resource capability or institutional prerequisites for quality forestry to take place; and *then* for certification to be possible. The availability of certification may, however, encourage individual companies and countries to invest further resources aimed at improving forest management.

Forest needs and certification's possible role

It seems that under certain circumstances, certification can be an effective market-based instrument. If certification is taken up, then what is its effect on forest problems? Table 6.1 summarises what this might be, related to the problems identified in Chapter 2 and the main needs outlined in Chapter 4. Certification is able to meet certain forest needs and has both direct and indirect effects on certain forest problems. These effects are discussed below. Our observations are based on experience of certification activities on the ground to date, and upon the performance of economic instruments in general.

Actual experience is currently quite limited and our main conclusions on certification's possible role are restricted to four main points:

1 **Certification as a market-based incentive applies to wood products which are sold in environmentally-conscious markets.**
   Certification is likely to be of most interest to LFMUs and to other companies in the supply chain providing products to these markets.

Table 6.1 *Forest problems and needs*

Forest problems	Forest needs							
	a increased stakeholder involvement	*b* policy/ law reform	*c* generating incentives	*d* new institutional roles	*e* improved efficiency of supply	*f* better international coordination	*g* new skills and capacities	*h* improved management control
deforestation	direct: requires local participation		indirect: product differentiation				indirect: sustainable management	
declining forest quality	direct: requires local participation		indirect: product differentiation				indirect: sustainable management	
diminishing biodiversity	indirect: requires local participation		indirect: product differentiation				indirect: sustainable management	
global warming			indirect: maintaining forest cover					
loss of cultural assets	direct: requires local involvement		indirect: product differentiation	direct: community involvement				
loss of local livelihoods	direct: requires local involvement		indirect: product differentiation	direct: community involvement	indirect: better access to markets			
over-exploitation	direct: requires local involvement		indirect: sustain-ability	indirect: public disclosure constraint	direct: long-term assurance		indirect: quality management	indirect: quality management
national inequity	indirect: requires local involvement			indirect: public disclosure				
international inequity			indirect: with international harmonization		indirect: better access to markets	indirect: disclosure, standards		
clashes of interest							indirect: participatory skills	
lack of trust between stakeholders	indirect: requires consensus			indirect: independent arbiter				
lack of political will				indirect: public accountability				
uncoordinated decision making			indirect: defined focus	indirect: process procedures		indirect: forcing issues	indirect: quality of procedures	indirect: better information
undervaluing forest resources			indirect: sustainability constraint					indirect: better information
overvaluing forest removal			indirect: sustainability constraint		indirect: better returns to foresters			indirect: better information
externalities not in prices					indirect: long-term returns			indirect: better information

direct: direct effect on forest problem (row)
indirect: indirect effect on forest problem (row)

	j harmonized data/ standards	*k* monitoring of multiple factors	*l* forest valuation	*m* more appropriate research	*n* better public reporting	*o* stable financial environment	*p* higher recovery of revenues	*q* funding incremental costs	*r* payments for 'global services'
					direct: better knowledge			*direct:* if market premium	
		indirect: ensure inclusion	*indirect:* sustainability constraint	*indirect:* identified needs	*direct:* better knowledge			*direct:* if market premium	
		indirect: ensures inclusion	*indirect:* sustainability constraint	*indirect:* identified needs	*direct:* better knowledge			*direct:* if market premium	
		indirect: ensures inclusion		*indirect:* identified needs					
		indirect: ensures inclusion		*indirect:* identified needs	*direct:* disclosure/ accountability				
		indirect: ensure inclusion		*indirect:* identified needs	*direct:* disclosure/ accountability				
			indirect: sustain-ability constraint	*indirect:* identified needs	*direct:* disclosure/ accountability				*indirect:* reduced environ-mental risk
	indirect: equal standards				*direct:* disclosure/ accountability				
	indirect: equal standards								
	indirect: equal standards	*indirect:* helps establish trade-off							
					direct: independent reporting				
					direct: independent reporting				
					direct: disclosure/ accountability			*indirect:* if market premium	
					direct: disclosure/ accountability				
					direct: disclosure/ accountability				

Certification in Context

Companies may either be interested in certification in order to exploit a perceived market opportunity; or use certification as a tool to reduce the environmental risk associated with selling wood and paper products in such markets.

2   **Certification acts at the forest level and is local in impact.** Certification has to rely on the addition of many individual forests before having a large direct impact. The first forests to become certified may well be those which are already being well managed, where the costs of preparing for successful certification are low. Certification will require considerable uptake before it can have a large-scale impact on the way that forestry operates; and on the forest problems which have been identified. It may need enabling regulatory support to do this; certification bodies alone may not be able to offer the institutional environment required.

3   **Much of the contribution from certification will be indirect.** Certification can work effectively as an indirect incentive to promote the activities required for sustainable forest management. Many of the needs identified in Table 6.1, for which certification may have a potential to contribute, fall into this category. In particular, the intense activity that currently surrounds certification is forcing and focusing attention both on distinct forest problems and on quality forest management. These are being addressed with a renewed vigour in many countries and internationally.

4   **Where certification can contribute, it is likely to form only part of the solution.** To be cost effective, certification often requires improvements in management quality. This implies investment and long-term commitment to training, systematic documentation layout, adherence to prescribed and documented operating procedures etc. Certification may be the catalyst to ensure management needs are addressed, but often other resources are still required to ensure that the need is satisfied. The policy and institutional environment also has to be conducive to quality forest management and to the practice of certification.

Our further observations on the potential role of certification are more tentative. These are summarized in the following paragraphs and follow the 'needs' listed as 'a' to 'r' in Table 6.1.

a   **Increased stakeholder involvement.** Certification requires consultation with stakeholders. This helps to ensure that adequate safeguards are in place for all goods and services from a particular LFMU; and, because stakeholder needs are addressed by the certification process, the extraction of forest goods and services – both for the market and to satisfy local livelihoods – becomes more sustainable.

b   **Policy/law reform.** Certification has no direct effect on national forest policy or legislation. Rather, adequate policy and legislation needs to be in place to assist certification. Pressure from certification initiatives may indirectly contribute to reform.

c **Generating incentives.** As a market-based instrument, certification is all about providing incentives. The extent to which certification as an incentive can address forest problems depends very much on how it affects the behaviour of commercial interests. This is likely to depend on its input cost, product price and market access effects, which have been discussed in the first part of this chapter.

Certification can also act as an incentive to harmonize forest management standards between countries; and to improve coordination of decision-making by defining a focus for sustainable forestry. This latter aspect may, in fact, be of more significance than is currently realized. The intense activity of environmental groups behind certification is forcing and focusing attention on distinct forest problems throughout the world. This, in turn, is helping to force the pace of other international initiatives designed to address forest problems. This debate, which is often formalized – as in the national working groups set up to define national certification programmes under the FSC – is also creating a climate of change for policy and legislative reform [see 'b' above].

d **New institutional roles.** As a market-based instrument, certification challenges the traditional role of government as majority stakeholder, regulator of forest activities and custodian of the national forest estate. But it also provides an orderly mechanism for other groups to play their legitimate roles, with the incentive to play these roles to a high standard and cost-effectively.

All stakeholders are involved in certification, results are publicly disclosed and stakeholders are accountable for their activities. The idea of government as a stakeholder becoming accountable through certification can be difficult for governments to accept. This is especially true where foreign organizations have been employed to undertake the certification.

Where certification becomes acceptable on a large enough scale, the process may appear to compete with the traditional regulatory role of governments. Such conflict is probably avoidable as, even with certification, certain government regulatory roles will be required. If certification is conducted within the context of an environmental management system, then the activities of the certification body would complement rather than duplicate other monitoring activities – including those undertaken by government. Chapter 7 includes a detailed description of environmental management systems.

By including all stakeholders and by providing an independent assessment of an LFMU's forestry activities, certification will necessarily challenge existing institutional structures; and assist in their development so as to better meet today's needs. At present, these structures are rarely supplying all the services required to address current forest problems.

e **Improved efficiency of supply.** By improving the efficiency and transparency of the supply chain, certification can reduce the number of intermediaries and thereby increase the proportion of the final sale price awarded to the forest owner. To do this, certification has to act in two ways. Firstly, certification should clearly identify the product's source to the consumer and retailer. Secondly, by requiring quality management, certification reduces both the transaction risk and the need for retailers to buy through traders who have traditionally absorbed this risk.

For an LFMU to be certified, certification has to be affordable. This requires the existence of inspectable and documented systems. By virtue of improved attention to such systems it often then follows that the management quality of the LFMU will be high. This indication of management quality provides greater confidence that wood will be delivered to specification and on time; thereby reducing perceived transaction risk. Buyers are more likely to buy directly from the forest owner.

Improved supply prospects can be of particular benefit to smaller forestry operations, in terms of providing direct market access and in obtaining better prices for wood products. This is especially true in situations where buyers are actively looking for certified sources.

f **Better international coordination.** It is unlikely that certification itself will improve the international coordination required to address many forest problems. Should certification gain international recognition, then better international coordination would conversely improve the effectiveness of certification. However, as mentioned earlier, the current activity and discussion which surrounds certification is having some effect in coordinating debate in many international fora, especially with regard to the need for forest monitoring, international accountability and harmonized standards for sustainable forest management.

g **New skills and capacities.** Sustainable management of the world's forests requires new skills for stakeholders in almost all forests, as well as new capacities for organizations involved in forestry.

Currently acceptable definitions of sustainable forest management are very different from those used ten or twenty years ago. Until quite recently, a main requirement of forests was their ability to sustain wood supplies. As a result, most professionals working in forestry today have been trained and have worked in an environment where performance is assessed on the basis of the quantity of wood produced. The current interest in certification and standards may, in fact, signal broad concern amongst some stakeholders regarding the professional skills and conduct of foresters in light of today's needs.

Almost everywhere a major effort is required in retraining to ensure that people entering forestry have skills which meet today's needs, and in adapting organizations to cope with the 'new forestry'. Such requirements often go against entrenched positions.

Certification develops new skills for the certification body in assessment; and for the LFMU in a range of long-term silvicultural and forest management activities, quality management and managing stakeholder participation. It may therefore contribute to foresters' professional development.

h Improved management control. This is probably one of the greatest benefits of certification. Transparent and efficiently functioning systems are almost a prerequisite for certification to be cost-effective. Good systems also provide accurate and timely information, which assists management in making better decisions and improving control over what is happening in the forest.

Chapter 7 includes a detailed description of the benefits to the LFMU from internalizing much of the certification process in the form of a quality management system.

i Permanent forest estate. Certification has no direct effect on the selection and demarcation of a national permanent forest estate. It is likely, however, that for certification to be of interest to forest users, a permanent forest estate must exist due to the long-term nature of the investments required to achieve sustainable forest management. Certification does not encourage the creation of new forests, as it focuses on the management of existing forests of any type.

j Harmonized data/standards. Certification programmes are harmonized where common external standards have been adopted; and accreditation allows for mutual recognition of certificates. Where such harmonization is achieved – and provided it is accompanied with safeguards that permit equality of access to the certification programmes – then certification can help to reduce problems of varying data definitions.

k Monitoring of multiple factors. Certification must take into account all goods and services required by stakeholders from an LFMU and the environmental and social impacts of their production. This ensures consideration of non-wood forest products. These can be of great importance to the livelihoods and culture of local people. In much of Eastern Europe, for example, maintaining access to forest fruits and mushrooms is an integral part of forest management. In much of India the forest is of more importance for cattle grazing by local people than it is for commercial timber production.

By forcing examination of multiple goods, services and impacts, certification analyses a complete picture of the forest. In this way, certification indirectly offers methodologies whereby multiple factors could be more efficiently and better monitored by national and international bodies responsible for forest condition.

l Forest valuation. Certification requires that the flow of goods and services from an LFMU should not exceed the LMFU's ability to provide the same level of goods and services in the future. Maintaining a constant value of goods and services in perpetuity ensures that the value of the forest's capital stock is not reduced.

As a result, insights are given into what total forest valuation might look like and the factors involved. Even a small area of certified forest can begin to provide such information. For this to happen, the results of certification must be expressed as far as possible in quantitative terms, or at least in clearly-defined physical terms and should be effectively communicated.

m   More appropriate research. The certification process focuses on specific forestry activities. The effect of these activities is assessed against the certification standards and decisions are made on the extent to which these standards are met. These decisions also point out gaps in existing activities. These gaps can be used to identify research needs. Experience to date suggests that, in humid tropical areas, a common need is cost-effective methods of undertaking multi-factor management inventories; and almost everywhere ways are needed to manage stakeholder participation.

Certification can help to define research needs more accurately. Increased precision in the identification of research needs directly relevant to the LFMU could enable a more efficient allocation of research resources.

n   Better public reporting. One of the core tenets of certification is the provision of independent statements on forest condition and status: the principle of third party verification. In this, certification has much in common with accountancy. Both depend on the accreditation rules, code of conduct and professionalism of the individuals involved. As with accountancy, the credibility of the certification process lies in the production of independent audit results. In order to ensure both transparency and accountability, these results should be publicly available. Progress in solving many forest problems is constrained by a lack of information on what is really happening in forests. Certification ought to be able to alleviate this both directly and indirectly.

Apart from the public reporting of activities in certified forests, certification can provide an impetus for better reporting of activities in other forests.

o   Stable financial environment. Certification is not expected to produce a more stable financial environment itself. Rather a more stable financial environment may encourage companies to make the necessary long term investments required for sustainable forest management; and this will assist in creating the conditions for certification to be a realistic commercial option in many instances.

Certification may also assist companies in raising new funds through rights issues and/or loans, by reducing the perceived environmental risk that such projects carry. Due to increasing public pressure, investments in natural forest exploitation carry particular risk.

p   Higher recovery of national forest revenues. Certification requires compliance with national legislation and regulation. However, because certification concerns individual forest areas, it will have an impact on the collection of forest revenues only if it becomes widespread. Where forest revenues are being avoided, certification may improve collection. In practice, organizations which volunteer for certification are unlikely to, knowingly, be avoiding the payment of forest revenues.

q **Funding extra costs incurred in achieving sustainability.**
Financing quality forestry is always difficult in a competitive market place; and there is an increasing level of environmental risk associated with investments in forestry. Certification can reduce both of these problems by assessing the sustainability of the forest resource and by ensuring the forest itself is well managed. As a result, certification may assist forest managers in raising funds and obtaining access to cheaper finance.

r **Payments for 'global services'.** Mobilizing the resources to provide payments for global services, such as forests' role in carbon sequestration and maintenance of biodiversity, will require large amounts of political will. Certification on a large scale could help by providing independent information on the status of forest areas and on the costs associated with quality forestry. Such information is, however, unlikely to be available in the immediate future.

Part 2

Certification in Practice

7. Design Issues

In the last chapter, we considered how certification might contribute to solving forest problems; and examined the effectiveness of certification as a market-based instrument. Where certification has an identified role, however, certification programmes need to be well-designed in order for this role to be effective. The LFMUs participating in certification programmes also need to be operating at a minimum level of 'quality' – in terms of efficiency, procedures and management – in order to benefit from the process. This chapter considers the design and participatory issues of certification within such a context.

If certification is both commercially attractive and practicable, it should be able to attract many subscribers, and be effective in improving forest management over large areas. It follows that the design considerations of a certification programme should satisfy these goals; as well as ensuring that the environmental level of performance set by the programme's standards is met.

Central to a successful certification programme must be:

- **Mechanisms which ensure credibility of the assessment process.** One of the best means for achieving this is effective accreditation of certification bodies. Credibility of the certificate is central to its commercial value; and credibility of the certificate depends almost entirely on credibility of the process by which certification bodies are accredited, and the professional standards to which such certifying bodies are expected to operate.

- **An LFMU structure which allows for cost-effective certification.** One of the best means by which this can be achieved is for LFMUs to implement an environmental management system [EMS]. For certification to be cost effective, the LFMU being certified must be easily assessable. That is, it must have a clear and documented management system in place. The level of documentation should be appropriate to the complexity and size of the forest being assessed.

Figure 7.1 and Box 7.1 show and describe, in general terms, how we recommend a certification programme should be organized; and the relations and main processes which govern its operation.

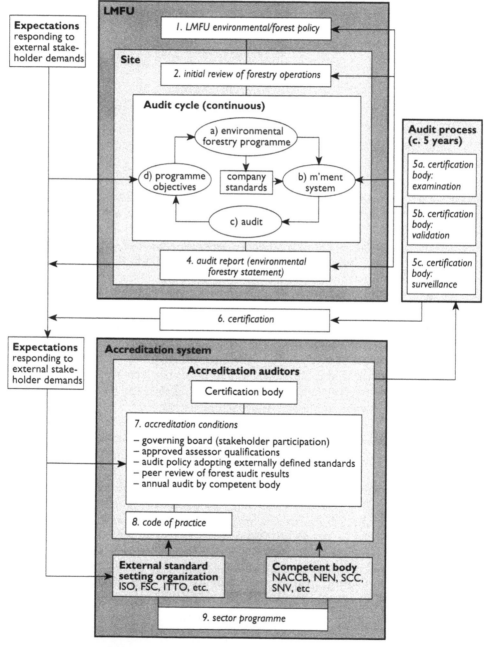

Figure 7.1 *Overview of a forest certification programme*

Box 7.1 *Overview of a forest certification programme*

Forest certification, which includes third party inspection and adequate public disclosure, is one way in which the LFMU can respond to calls for accountability [paragraph numbers below relate to numbers in Figure 7.1].

1   An LFMU that wishes to have its forestry activities certified prepares an *environmental forest policy*, clearly stating its objectives. The policy should be endorsed by the senior forest manager.
2   An *initial review of the LFMU's forestry activities* is undertaken by the certification body. This examines the extent to which the LFMU has procedures in place which ensure that stakeholder concerns are addressed as part of the management structure; and are fully incorporated into all activities.
3   The operation implements *planning and management procedures* which ensure that all environmental issues identified in the policy are addressed. The process is continuous and includes: [a] internal development of an *environmental forestry programme* which includes standards of forest practice; [b] development of an *environmental management system*, to ensure that the environmental forestry programme is implemented and forest practice standards are adhered to; [c] a *regular internal assessment procedure* to measure compliance of forestry activities with the documented procedures described in the management system; and [d] *use of the assessment process* to identify the degree to which programme objectives have been met and to improve further the environmental forestry programme.
4   An *internal assessment report* is produced, which the LFMU can use to communicate the environmental impact of their forestry activities.
5   A *third party independent assessment* is carried out that has three main components: [a] an *examination by the certification body of appropriate documentation* in order to objectively assess that management systems can meet required environmental objectives; [b] a *sample field validation* by the certification body, to assess whether field forestry practices comply with documented stipulations; and [c] where required, *surveillance visits between assessments* to ensure continued compliance with environmental objectives.
6   Following a successful third party assessment, the LFMU may *apply for certification* that its forestry practices meet specified standards.
7   Accreditation conditions determine the credibility of the third party assessment and certification process. The *independent certification body should be accredited* according to conditions which are specified by an external forestry standard-setting organization – such as the FSC; and also by a national competent body where possible and appropriate.
8   Where accreditation to an external forestry standard-setting organization may not formally be available, certification bodies can follow their own *internal code of forestry practice, backed up by existing external standards* as appropriate. Procedures used by the certification body should be overseen by a governing board; and assessment results monitored by an independent peer review board.
9   A *sector programme* can be established, with participation of both the national competent body and the external forestry standard-setting organization.

Certification in Practice

The design issues to consider in certification can be divided into three groups:

- the importance of environmental management systems at the LFMU level;
- accreditation; and
- the certification body.

The importance of environmental management systems at the LFMU level

Often, LFMUs have undertaken or commissioned assessments of their environmental performance. However, these alone cannot provide the assurance that an LFMU will *continue* to satisfy the environmental and social standards required. To be meaningful, such assessments need to be conducted within a structured management system. This is also true of certification and, where applied, is one of the ways in which certification can contribute to, and become part of, effective management. In addition, poorly-organized and weak management systems make certification difficult and increase its cost.

For these reasons, a forestry organization preparing for certification should consider implementation of an environmental management system [EMS] such as that specified in the British Standard BS 7750 [see Box 7.2 for a summary]. EMS standards have recently been developed by a number of countries [see Box 7.3] and ISO is also developing an international EMS standard. The ISO standard is to be called ISO 14000 and will be analogous to ISO 9000[1] – an international standard for quality management systems.

ISO 9000 contains generic guidelines for *quality systems*, as opposed to technical specifications for a particular *product*. The standard is internationally accepted and recognised as a mark of quality management – ISO likes to present ISO 9000 as an 'international visa' of quality. Since its launch in 1987, over 50 000 companies around the world have been certified. More than 70 countries have adopted ISO 9000 as their national standard for quality systems. ISO 9000 is becoming a fundamental contractual requirement for doing business with governments and trading blocs and within many major economic sectors.

In being a valid precursor for development of an EMS, ISO 9000 can also be considered as a starting point for LFMUs contemplating certification; and in addition ISO 9000 has value in its own right.

In principle, an EMS is no different from any other sort of management system. It contains the same components. It aims to identify and improve processes

---

1 ISO 9000 is the international standard equivalent to EN 29000 at the European level and the British standard BS 5750. The British standard BS 7750 is compatible with BS 5750 as they take parallel approaches. Organizations operating BS 5750 are able to extend their management systems to meet the requirements of BS 7750. However, operation to BS 5750 is not a prerequisite for operation to BS 7750. It is expected that development of the ISO standard for environmental management systems [ISO 14000] will take a similar approach. Organizations already certified to ISO 9000 will be able to extend their management systems to meet the requirements of ISO 14000.

Box 7.2 *BS 7750: a specification for environmental management systems (EMS)*

BS 7750 is designed to enable any organization to establish an effective management system as a foundation for sound environmental performance. The standard shares common management system principles with ISO 9000. Organizations can use an existing management system, developed in conformity with ISO 9000 as a basis for environmental management. BS 7750 does not establish absolute requirements for environmental performance beyond compliance with applicable legislation and regulations, and a commitment to continual improvement. Thus, two organizations carrying out similar activities but having different environmental performance may both comply with its requirements. It should however be noted that BS 7750 does require the organization's environmental policy and objectives to be publicly available.

The philosophy behind BS 7750 is that environmental reviews and audits *on their own* cannot provide an organization with the assurance that its performance not only meets, but will continue to meet, environmental requirements. To be effective, environmental audits and reviews need to be undertaken *within a structured management system*, integrated with overall management activity and addressing significant environmental effects. In BS 7750, environmental audits assess both the effectiveness of the EMS and the achievement of the environmental objectives and targets. Environmental reviews check the continuing relevance of the environmental policy, update the evaluation of environmental effects, and check the efficacy of audits and follow-up actions.

The requirements of BS 7750 are:

1 the documented EMS shall take account of any pertinent code of practice to which it subscribes;
2 a documented environmental policy;
3 clear definition of all organizational responsibilities and authorities;
4 procedures to ensure that contractors are made aware of all environmental requirements;
5 procedures for receiving and responding to stakeholder views;
6 procedures for identifying and evaluating direct and indirect environmental effects of its activities;
7 procedures to specify environmental objectives and consequent targets;
8 establishment and maintenance of a programme to achieve the objectives and targets;
9 a documented system including an environmental management manual;
10 procedures which ensure that activities with a significant environmental impact take place under controlled conditions;
11 procedures for verification of compliance with specified requirements – including targets;
12 procedures for initiating investigation and corrective action in the event of non-compliance;
13 records which demonstrate performance of the EMS; and
14 undertaking regular and periodic audits and reviews.

and relies on good communication to be effective.

The concept of EMS is applicable to any LMFU, regardless of size, type or level of sophistication. To be successful across a wide spectrum of LMFUs, an EMS must be implemented using a *process* rather than a functional approach.

Implementation of an EMS should also assist in achieving greater competitiveness and consistently delivering environmental and social assurances to stateholders. True benefits – such as more efficient processes, lower costs, customer commitment, higher profits, more motivated employees – can also only be achieved by a process-based approach to EMS.

This requires recognition that everything which happens in an LFMU – whatever its size, makeup or sophistication – does so by virtue of someone doing something. People and processes unite everyone in an LFMU regardless of function, department, technical or professional expertise. It is only when both the LFMU and certification body come to grips with this reality that an effective EMS can be designed to meet the requirements of quality forestry.

Building an EMS based on the way an LFMU actually operates is both simpler and avoids production of a paper bureaucracy.

An EMS emphasizes prevention rather than cure; and the continuous improvement of environmental performance. A well-functioning EMS builds environmental and social considerations into all parts of an LFMU's activity. It enables the LMFU to provide objective evidence that policies, objectives and targets meet stakeholder expectations; and give evidence of legislative and regulatory compliance. The relationship between an EMS and certification is illustrated in Figure 7.2.

Specific benefits associated with an EMS could include:

*   preparation for certification;
*   meeting client environmental and social expectations;
*   demonstration of due diligence;
*   satisfying total investor and lender criteria, leading to an improved ease of raising capital;
*   limitation of liability and environmental risk, resulting in avoidance of prosecution and ability to obtain insurance at reasonable cost;
*   enhanced image and market share;
*   cost control including conservation of input materials and energy;
*   improved long-term wood supply potential and enhanced access to forest areas for harvesting;
*   development of improved processes and techniques;
*   improved LFMU-to-government relations;
*   maintaining good relations with local communities and the public in general; and
*   improved forest environment and human welfare.

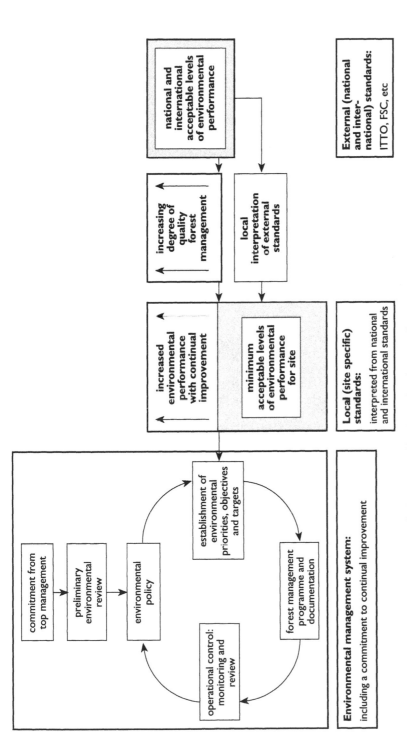

Figure 7.2 *Forest certification and environmental management systems*

Box 7.3 *Development of environmental management systems*

In 1992 the British Standards Institute [BSI] launched its standard BS 7750 'Specification for Environmental Management Systems' (Box 7.2). This has been followed by similar and compatible initiatives in other countries:

*   in France the *Association Francaise de Normalisation* [AFNOR] launched X 30-200 'Environmental Management System' in April 1993;
*   in Eire the National Standards Authority launched IS 310 'Environmental Management Systems – Guiding Principles and Requirements' in 1994;
*   in the United States the American Society for Quality Control and the American National Standards Institute launched ANSI/ASQC E4-1993 'Quality Systems Requirements for Environmental Programmes' in 1993; and
*   in South Africa the South African Bureau of Standards launched SABS 0251:1993 'Code of Practice, Environmental Management Systems' in 1993.

In June 1993 the European Council issued Regulation No. 1836/93 allowing voluntary participation by companies in the industrial sector in a Community Eco-Management and Assessment Scheme [EMAS]. EMAS is due to come into effect during 1995 and covers:

*   requirements for a company's environmental policy, objectives and programmes, environmental management systems [similar to BS 7750] and good management practices;
*   requirements concerning assessment, its methodology, coverage and frequency;
*   accreditation criteria for certification bodies and their functions and actions during an assessment; and
*   institutional aspects to be covered by Member States

In response to this proliferating EMS activity the International Standards Organization [ISO] began the process of developing an international EMS specification. This specification is being developed by a technical committee [ISO/TC207/SC1/WG2] and has the current number ISO 14000:199X. Guidelines are currently in draft form and are expected to form the basis for an eventual international standard for EMS which will be analogous to ISO 9000.

An EMS provides the starting point and tools with which LFMUs can effectively balance and integrate commercial interests with those of society and the environment. It therefore helps such LFMUs to prepare for certification.

There are four key principles in an EMS:

1   definition of purpose: focus on what needs to be done – and establishment of a plan;
2   establishment of commitment: by everyone in the LFMU;
3   provision of adequate resources: so that the organization is capable of implementing the EMS; and

4 continuous improvement:- so that the forestry organization is continuously evaluating, learning and improving.

Implementation of these principles requires six main actions:

1 the commitment of top management;
2 completion of an initial review or baseline survey;
3 setting an environmental policy and objectives;
4 developing an environmental management programme;
5 a system of operational control and monitoring; and
6 an allowance for periodic review and assessment.

- **Top management commitment** is the first step in developing an EMS. Without top-level commitment, the system is almost certain to fail.
- **An initial environmental review [baseline survey]** is required to establish the current position of the LFMU with regard to the certification standards and the LFMU's environmental policy. The review should take account of:
  - legislative and regulatory requirements;
  - identification and evaluation of environmental and social effects of current activities;
  - establishment of environmental priorities (including species, communities and habitats which require active [as opposed to passive] conservation) and social priorities (sustainability of forest communities; fair and effective decision making);
  - examination of environmental and social effects from existing activities and procedures, including identification of areas where improvements are required;
  - assessment of procedures for stakeholder participation;
  - assessment of internal monitoring and feedback procedures;
  - identification of skill and other resource shortages including details of additional expertise, training and other resources required;
  - adequacy of recording, documentation and its control;
  - details on opportunities for competitive advantage; and
  - assessment of environmental risks and liabilities.

In all cases, particular consideration should be given to possible non-conforming activities; clearly distinguishing them from acceptable activities.

  An initial environmental review should be undertaken each time the forest area is increased, with the acquisition of either additional standing forest or additional land for tree planting. The level of detail required by such reviews will generally be less than the initial environmental review, but will depend on the existence of additional significant environmental effects.

  The scope of the review should cover: geology, soils and topography;

previous land use – for plantations; hydrology – especially downstream effects; fauna and flora; climate; archaeology and cultural items; socio-economic conditions – internal and external to the operation; and public concerns. The review should not be confined within the forest boundary alone.

- **Setting an environmental policy:** this should detail the LFMU's overall aims and principles of action with respect to the environment. Statements should be specific and clear. The environmental policy should provide for compliance with accepted external standards and all relevant environmental regulatory requirements, and express clear commitments aimed at the continual improvement of performance. In addition the policy should highlight top management commitment by being approved by the board of directors or other governing body and signed by the chief executive officer/director. It should state the LFMU's commitment to meet the reasonable expectations of its stakeholders; and be effectively communicated to both internal and external stakeholders – and also be understandable to LMFU personnel, contractor personnel and the public.

- **Development of an environmental management programme** incorporates both the findings of the initial review and the principles set out in the environmental policy. An environmental programme specific to the LFMU is then drawn up. The environmental programme should:
  - *identify and evaluate the risks associated with all activities*, to assist in prioritizing environmental and social effects. This is an ongoing process that identifies the current and future impact of LFMU activities on the forest;
  - *define objectives and specific targets* Proper definition of objectives and targets relies on correct identification and assessment of environmental and social effects as part of the initial review. Defined objectives and targets should also be consistent with the environmental policy.

Objectives, which are quantified wherever possible, are the overall aims that the LFMU sets itself to achieve. They should reflect the identified environmental and social effects and the resulting priorities within the scope of the external standard.

Targets are the detailed performance requirements set in order to ensure that objectives are met. They should not be measurable performance requirements which an LFMU sets out to attain over a specified period of time. Regular internal monitoring to measure the organization's perfor-mance against the targets allows for re-examination of effects, their priorities and objectives; and provides for the basis for continuous improvement. Both long-term [five years] and short-term [annual] objectives and targets should be set. For example, the policy could be a commitment to 'conserve biologi-cal diversity'; the objective 'to ensure that forest operations maintain the ecological functions and integrity of the forest'; and the target 'to maintain a minimum area of undisturbed forest equal to 10 per cent of the forest area'. From the target, an action plan should be defined, with responsibility for its implementation identified.

Objectives should allow for environmental variation between different forests but be consistent with the same external standards. Together, objectives and targets define the operational standards expected from the LFMU. Figure 7.3 illustrates this linkage between policy, objectives and targets.

Where there is consistent failure to meet the required operational standards, the operating procedures are examined to identify areas where improvement is required. Such improvement may require considerable change involving new operating procedures and resources. It should be noted that effective definition of objectives and targets relies on correct identification of environmental effects:

— *ensure capability to act in support of environmental commitments.*

The resources needed by the LFMU will evolve constantly in response

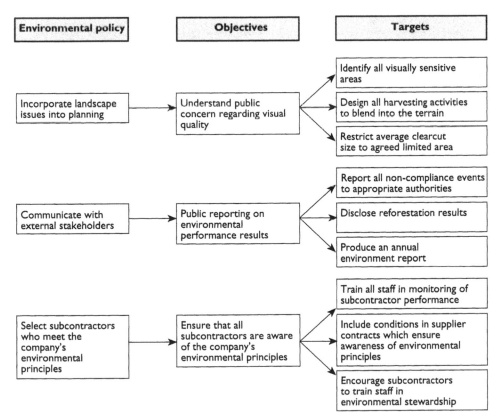

Figure 7.3 *Relationship between 'policy', 'objectives' and 'targets' within an EMS: some examples*

to changing stakeholder requirements, a dynamic business environment and the process of continuous improvement. This will require appropriate human, physical [eg facilities and equipment] and financial resources. In particular, responsible staff need to have suitable knowledge and skills, and where these are not evident suitable programmes for training and skill development need to be in place;

— *emergency planning is required to respond to unexpected incidents –* such as an overturned diesel tanker lorry with resultant spill into drainage works. Emergency plans should include: defined roles and responsibilities; details of emergency procedures; communication procedures; training plans and testing for effectiveness;

— *establish internal and external communication and reporting processes.* These should communicate in a credible way with internal and external stakeholders concerned with the organization's environmental effects and performance.

Good information management is an essential part of an effective EMS. The environmental management plan should clearly define and document operational processes and procedures. These should be updated as required. The preparation and use of EMS documentation helps to evaluate the system, improve environmental performance, and ensure that employees are aware of what is required to achieve the LFMU's environmental objectives.

It is often desirable to maintain a summary of EMS documentation which would include the environmental policy, objectives and targets; a description of the means by which objectives and targets will be met; key roles and responsibilities; and direction to related documentation.

In particular, the LFMU should also establish and maintain procedures for controlling relevant documentation to facilitate review and revision, and for removing or marking all obsolete documents.

- **A system of operational control and monitoring** is required for implementing the environmental management programme. The establishment and maintenance of controls [internal standards – objectives and targets, procedures, instructions, programmes] ensures that the level of environmental performance is consistent with the LFMU's policies, plans, objectives and targets. Controlled work programmes are established through the development of operating procedures or work instructions. Operational controls are established to guide performance.

Operational control and monitoring is essential for an LFMU to evaluate, learn and improve; and to continuously improve. Continuous improvement is achieved by improving processes within the LFMU and is sustained by training and learning.

Effective monitoring requires standards. Internal standards which are relevant to the particular forest in question are developed as part of the EMS in terms of objectives and targets (see above). These should be consistent with the external standards set by a certification programme.

Monitoring can be expensive. The objectives of the monitoring programme need to be clearly defined and focused on environmental priorities, where such a need demonstrably exists. A monitoring programme focused in this way is more likely to produce information in a concise form; and is more likely to be used by the forest manager.

- **Periodic reviews and assessments** can be carried out by the LFMU's personnel, and/or by external parties. In either event, the assessor should be independent of the party being assessed, and properly trained to carry out assessments objectively and effectively. External assessing will be a requirement of the certification programme.

An assessment may be of a particular aspect of the EMS. From time to time reviews of the entire EMS should be undertaken – testing management processes as well as whether objectives and targets are being met. In this case, the scope of the assessment should include the entire LFMU and its activities, products and services. The assessment should not be confined to the forest area. Specifically an assessment will:
  - review environmental objectives, targets and performances;
  - review findings from previous assessments and any monitoring activities;
  - assess the continuing suitability of the EMS policy and the need for changes in the light of: changing legislation; changing expectations and requirements of stakeholders; changes in the products or activities of the LFMU; advances in science and technology – including latest accepted 'best forestry practices'; lessons learned from environmental problems in other similar forest situations; and market preferences.

The findings, conclusions and recommendations reached as a result of the assessment and review of the EMS should be documented, and the necessary changes implemented by the LFMU's senior management. Management should ensure that there is continuous follow-up on the assessment and review findings.

In the EMS context independent third party assessors complement other monitoring activities – including those undertaken by government agencies. This can be demonstrated quite clearly if one views the processes within a typical national forest sector through an EMS-type framework as illustrated in Figure 7.4. The usual role of government forestry departments is to interpret national and local government forest legislation in the form of a policy and associated documents. Preliminary reviews are often commissioned to assist in the formulation of objectives and targets (standards) and management programmes. At an operational level, objectives are built into a forestry code of practice or environmental guidelines for forestry operations. Management programmes, including targets, are then defined by state-owned or privately-owned forestry operations, in order to satisfy the prescribed objectives. The national forest service or other regulatory authority then, frequently, monitors the effectiveness with which the

Certification in Practice

management programmes are implemented and the extent to which the objectives and targets are complied with. The results of the monitoring exercise may in turn encourage improvements to national forest policy. Even at a national level, continuous improvement is also possible where all links in the cycle are in place and functioning well.

At a national level the role of the assessor is to provide central and local government with an assessment as to the effectiveness with which the forestry sector management system is functioning. The assessment results can be used to improve the quality of the systems being used. In this way the environmental assessment becomes an integral and necessary complement to effective administration of the forestry sector. In most countries, financial assessments are undertaken which cover all aspects of the forestry sector. Such assessments are already considered to be an integral and necessary part of activities. Despite this, many governments view the possibility of independent environmental assessments with concern.

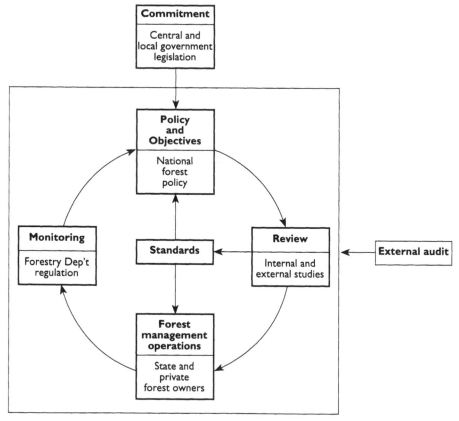

Figure 7.4 *A typical national forest sector seen through an EMS-type framework*

Accreditation

Accreditation of certification bodies becomes particularly important when extensive professional judgement is necessary for the local interpretation and application of standards. This will almost always be the case in forestry.

In general three principles should be followed:

1   Accreditation should focus on the *procedures which certification bodies* adopt in their work, in the same way that the certification of forest organizations starts by looking at the management systems in place. Certification bodies should have systems in place which enable them to proceed with the diligence required; and, importantly, to replicate the same assessment over the same area of forest and arrive at the same result. Standards are always externally set. Nevertheless the certification body should express its own code of professional conduct and/or have an environmental policy.
2   Accreditation should examine the *structure of the certification process*. The certification body should have in place a policy document or description of its certification programme which allows for a clear understanding of the steps and controls allowed for in the certification process.
3   Accreditation should look at the *qualifications of the individuals to be employed* by certification bodies in undertaking the assessments. Individuals should have adequate levels of professional qualification plus specific training and experience in assessment work. Assessing is not the same as consultancy and specific expertise is required. Assessors involved in ISO 9000 certification are required to undertake specialized lead assessor training and to have participated in a minimum number of assessments. Similar requirements should be defined for forestry.

The structure of a certification programme needs to be governed by various general guidelines and criteria in order to earn consumer acceptability and achieve market credibility. This is true for all products where certification is required and accreditation is designed to satisfy these requirements. *Accreditation effectively licences or franchises certification bodies to operate, provided that they follow clearly defined and accepted rules.* It follows that accreditation forms the core of any certification programme.

For these reasons, there are several general guidelines and criteria to govern the structure of certification programmes at national and international levels. Initiatives have been made by ISO, the joint European Standards Institution [CEN/CENELEC] and various national standards bodies.

Certification bodies wishing to become accredited for forest certification should be structured according to EN 45012 'General criteria for certification bodies operating Quality System certification'. EN 45012 is directly applicable in all CEN/CENELEC countries.[2] EN standards can also be used, and many

Certification in Practice

2 Austria, Belgium, Denmark, Finland, France, Germany, Greece, Ireland, Italy, Luxembourg, Netherlands, Norway, Portugal, Spain, Sweden, Switzerland and the United Kingdom.

actually are, in non-CEN/CENELEC countries. For example, in Indonesia the norm DSN 16 is directly equivalent to EN 45012.

Accreditation must play a large part in providing the certification programme with market credibility. Because of the market-based nature of certification and labelling programmes, their success depends upon market credibility and the ability to earn consumer confidence. This in turn requires that two key aspects of such programmes are addressed by the accreditation process:

1   **The structure of the certification programme must be transparent and the independence of the label sacrosanct.** The institutional framework of the certification programme must include a minimum number of checks and balances which avoid conflicts of interest between the different participants.
2   **The attributes of the certification programme and the label are clearly promoted to consumers.** An active and well thought out information and advisory programme must accompany a certification programme for it to be successful. The consumer must be provided with adequate information upon which to base purchasing decisions.

The certification body

As mentioned above, EN 45012 sets out broadly how a certification body should be structured. This is illustrated in Figure 7.5.

Accredited certification bodies can be either an organization or, with more limited accreditation scope, individuals. In all instances, the certification body should demonstrate competence in forestry practices and have personnel who are qualified, trained and experienced in:

*   environmental assessment methodologies;
*   management information systems and processes;
*   environmental forestry issues;
*   relevant legislation and standards; and
*   forestry practice.

In all cases the certification body must be independent, impartial and able to demonstrate that its organization and personnel are free from any commercial, financial or other pressures which might influence its verification activities and judgement or endanger its trust. Certification bodies complying with EN 45012 will have met these conditions.

In addition to satisfying accreditation rules, certification bodies must, as a minimum, have:

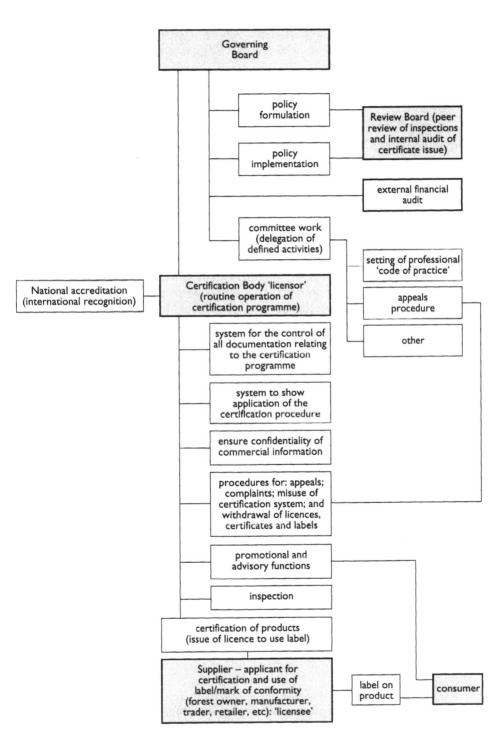

Figure 7.5 *General structure of a certification programme*

- documented procedures and methodologies to meet: assessment and verification requirements; quality control mechanisms; and confidentiality provisions; and
- publicly available: information detailing structures and responsibilities in its organization; and a statement of legal status, ownership and funding sources.

Certification bodies have two clear roles: to *examine* and test the documented management system of the LFMU; and to *validate* that the site-specific standards being worked to by the LFMU, and which are documented in its management system, are being satisfied in the field. Further details on the certification process are described in Chapter 8.

The activities of the certification body should be monitored by a governing board which is responsible for its performance. The basic functions of the governing board are policy formulation; overseeing policy implementation; and setting up technical committees as required, to which activities are delegated. For example, these would be required for the verification of assessment procedures.

The governing board should be representative of all stakeholders. No single interest should predominate. The structure of the governing board should safeguard impartiality and enable participation from all parties concerned.

As part of the assessment and control procedures required, it may be considered worthwhile to install a technical review board capable of controlling and verifying the quality of inspection procedures and reports. The review board could also monitor the issuing of certificates. In addition, an external financial assessment should be undertaken annually.

The certification body must be impartial and have a certain minimum number of attributes:

- its policy should provide for a mechanism by which members of the governing board are chosen;
- its personnel should operate under a senior executive responsible to the governing board, and should be independent from those who have a direct commercial interest in the certification to be carried out;
- its organization should clearly show the relationship between assessment and certification functions;
- its documentation should identify its legal status; describe its certification system and the rules whereby certificates are granted; and include a procedures manual;
- documentary systems should be in place to allow for adequate control of the certification system; and to record how each certification procedure was applied;
- adequate arrangements should be in place to ensure confidentiality of information obtained in the course of certification activities;
- details of the certification programme and certified products should be

promoted in publications; and
* procedures should be in place for appeals; complaints; misuse of the certification system; and withdrawal of licences, certificates and labels.

An appeals procedure for all types of complaint would, in the last resort, resolve disputes under the governing law. In practice, disputes should be few and far between, provided that the certification programme involves all interested parties in a true consensus; and that the certification body provides sufficient information on its activities. Disputes that do occur should be settled internally with unanimous agreement of the governing board considered as binding – or some other form of internal arbitration procedure. Only in the event that agreement cannot be reached by the governing board would legal measures be required.

8. How Certification Works in Practice

The previous chapters have shown where certification can contribute to solving forest problems; and discussed where certification might fit in and how it should be used. Chapter 7 illustrated the various components of a certification programme and how they should be structured to get the best out of the certification process. This chapter describes how the certification process works. That is, how the forest manager actually goes about getting a forest certified.

Each of the four active certification programmes described in Chapter 11 of Part IV, follow slightly different procedures. However, the main steps are common to all. These usually involve some form of application; a pre-audit site visit and discussion; the audit itself; peer review of results; and issuing of the certificate. Most of the difference between the programmes is concerned with detail and emphasis. The differences that do remain are being ironed out as practical experience is forging a convergence of procedures towards that which is most feasible.

In this chapter we present a generic approach that we recommend for the practice of certification. The sequence of events follows those used in the assessment of quality and environmental management systems. Some certification bodies now have over fifteen years' experience in assessing quality systems and have developed procedures which can easily be adapted to assessing quality forestry. The current forest certification programmes include many of the elements suggested in the following recommended approach.

Prior to application the forest manager needs to complete a process of self-assessment to determine if the forestry operation in question is ready for certification. If the owner has already implemented an EMS – as outlined and recommended in Chapter 7 – this would have already taken place. In the absence of an EMS a number of questions should be asked. These are illustrated in the form of a decision tree in Figure 8.1.

There are four main steps to consider:

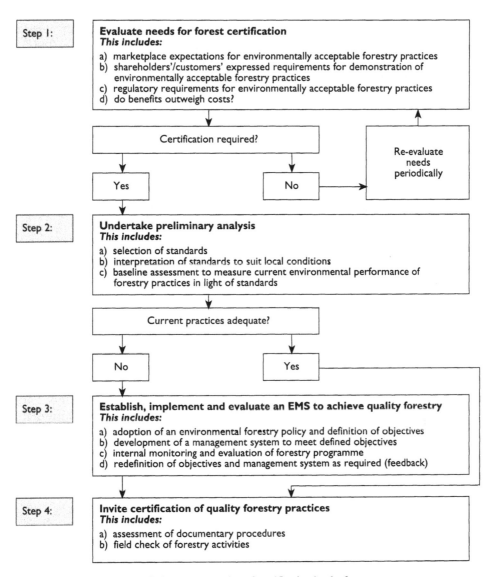

Figure 8.1 *Implementation of certification by the forest owner*

- **Step 1 – evaluation of need for forest certification.** The forest manager should be sure that demand for certification is real; that certification will confer a competitive advantage; and, in particular, that the benefit from certification will outweigh its cost.
- **Step 2 – undertake a preliminary analysis.** If a decision is made to go ahead with certification the owner should first prepare by undertaking a preliminary analysis. This involves the selection of standards and their

local interpretation; followed by a baseline assessment of current activities to measure their adequacy against the selected standards. If possible, the preliminary analysis will also identify and prioritize the environmental and social effects of the forestry operation. This will help to confirm the adequacy of objectives and targets to the assessor during assessment.

Use of the questions in Box 8.1 can provide the forest manager with a quick indication of whether more work is required to satisfy the demands of the certification process; and in which area of activity.

- **Step 3 – establish, implement and evaluate an EMS to achieve quality forestry.** Where inadequacies have been identified as part of Step 2, modifications will be required. These may be minor or major – perhaps even requiring a total rethink as to how the forestry activities are undertaken. In all cases it is recommended that modifications are made within the context of an EMS. This allows for clear and feasible objectives and targets to be set; procedures to be defined in order to meet set targets; and a programme of internal monitoring to measure progress.

  Such an approach also provides an auditable framework for the assessor during certification and is likely to accelerate the certification process. More importantly, a strong EMS is focused on future improvement. It is more likely to provide an assurance that the certified LFMU will continue to meet the demands of the certification programme.

- **Step 4 – Invite certification of quality forestry practices.** The bulk of the certification process takes place once the operation is well prepared and success likely to be achieved. However, certification is not an end of pipe, 'black' and 'white', 'no' or 'yes' process. In reality, the forest manager will have started a process of dialogue and understanding with the certification body from the beginning. In this way, award of the certificate is the culmination of a process which has resulted in improved environmental and social management of the forest in question; and – it is hoped – improved operating performance of the LFMU.

  Neither is the certification report simply a statement of whether a particular forest operation has complied with the standards in question. Recommendations are made for further improvement and actions requested to further strengthen management in problem areas. Usually time limits for the completion of specific activities or the attainment of set targets are agreed upon between the certification body and the forest manager.

Figure 8.2 illustrates the main steps of the certification process, identifying the responsibilities of the certification body and those of the forest owner or operator. These are described in more detail below.

Box 8.1 *Indicators to assist LFMU management in assessing their readiness for certification*

Policy and commitment:
- Is there a policy committing the organization to minimize the negative social and environmental effects of its activities?
- Is this policy endorsed by senior management and owners?
- Has this policy been communicated to all staff?

External regulation:
- Is it possible to easily and credibly demonstrate compliance with relevant legislation?
- Does the LFMU have a good record of legislative compliance?

Management:
- Are site-specific social and environmental objectives, targets and procedures integrated into the management system?
- Does a written management plan exist?
- Is there an active programme to ensure that all staff are adequately trained for the job?
- Does a monitoring programme exist for LFMU and contractor activities?

Optimum operations:
- Is there demonstrable evidence that the yield of forest goods and services is sustainable in the long-term?

Social effects:
- Have local people affected by the LFMU's activities been identified?
- Has a systematic appraisal been undertaken of the social effects of the LFMU's activities?
- Does a process exist for consulting with local people and acting upon their legitimate concerns?
- Can the LFMU demonstrate its right to custody over the forest resource?
- Are worker rights to organize for voluntary negotiations recognized?

Environmental effects:
- Has a systematic appraisal been undertaken of the environmental effects of the LMFU's activities?
- Do operating procedures exist which address environmental effects?
- Are responsibilities for environmental issues defined?
- Are environmental issues defined in agreements with contractors?
- Are WHO Type 1A or 1B pesticides or chlorinated hydrocarbons excluded?

If the answer to all these questions is 'yes' it may be possible to proceed directly with certification. If the answer to one or more of these questions is 'no' there is probably a need to implement elements of Step 3 [see main text] before proceeding with certification.

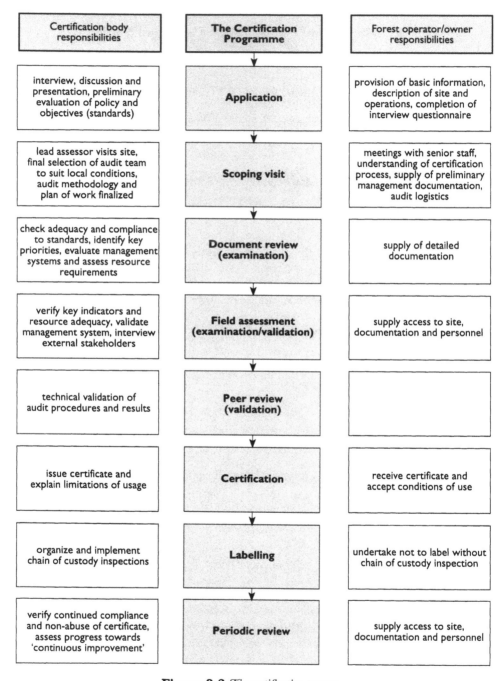

Certification body responsibilities	The Certification Programme	Forest operator/owner responsibilities
interview, discussion and presentation, preliminary evaluation of policy and objectives (standards)	**Application**	provision of basic information, description of site and operations, completion of interview questionnaire
lead assessor visits site, final selection of audit team to suit local conditions, audit methodology and plan of work finalized	**Scoping visit**	meetings with senior staff, understanding of certification process, supply of preliminary management documentation, audit logistics
check adequacy and compliance to standards, identify key priorities, evaluate management systems and assess resource requirements	**Document review (examination)**	supply of detailed documentation
verify key indicators and resource adequacy, validate management system, interview external stakeholders	**Field assessment (examination/validation)**	supply access to site, documentation and personnel
technical validation of audit procedures and results	**Peer review (validation)**	
issue certificate and explain limitations of usage	**Certification**	receive certificate and accept conditions of use
organize and implement chain of custody inspections	**Labelling**	undertake not to label without chain of custody inspection
verify continued compliance and non-abuse of certificate, assess progress towards 'continuous improvement'	**Periodic review**	supply access to site, documentation and personnel

Figure 8.2 *The certification process*

Certification in Practice

Selection of the appropriate forest area

Here we pause to consider more carefully the forest area appropriate for certification. In the introduction we defined this area as the LFMU, and suggested that it should be covered by a common management system. We also stated that selection of an appropriate LFMU is an important part of certification. Experience to date indicates that there are economies of scale in certification and that small areas proportionately cost much more to certify than larger areas. Certification becomes more cost-effective where the LFMU is in excess of about 100 ha of highly productive forest, or 10 000 ha of forest of low productivity.

Alternatively one could ask: at what point is a forest *too small* for certification? What interest would a farm woodland owner, producing a few logs of commercial value every twenty years, have in certification? How should certification proceed in areas where industry is buying wood from many small forest owners? How should certification be organized in areas of communal landholding; or where owners have multiple and small disparate forest blocks within a mixed landscape?

All these questions concern decisions on selecting an appropriate forest area. In making these decisions one should not allow the objective to be obscured by debate over detail. Certification is designed to work as an *incentive* whose objective is to improve forest management.

As a result, certification should operate in accordance with the level at which the incentive 'signal' is most efficiently received. This will usually be the forest owner. However, in many areas, especially where there are many owners with small forests, it will be necessary to divorce the issue of individual ownership from the certification of a forest area. This, in turn, requires a coordination of owner activities – such as through associations, cooperatives, industry supplier groups and/or groupings that follow existing administrative boundaries. These groups have to contain effective mechanisms which can pass on incentive signals to individual members; and to accept participation of individual members in a common management system.

Certification bodies should focus their attention on levels where common management systems already exist. Such systems may, initially, be incomplete and require further development in order to satisfy the requirements of certification. For example, a cooperative of forest owners established for the primary reason of selling wood is unlikely to have agreed forest management practices in place. They will need to bring together a strong and appropriate management system, which can be readily assessed.

Certification of multiple small woodland areas

Consider, as an example, a situation where industry is procuring wood from many small owners. Under such conditions – which are not at all uncommon – certification would need to be structured to satisfy three perspectives:

1 From the small woodland owner's perspective:
– there is no incentive to volunteer for certification as, typically, for many years there is no activity in the forest;
– the cost of certification per hectare and per m3 of wood sold is relatively high.

2 From industry's perspective:
– wood procurement 'assumes' a steady supply of wood which is either bought at roadside or delivered. As a result, industry often has little reason to be aware of forest management conditions;
– controlling the cost of raw materials requires flexible wood procurement – ie not necessarily limited to certified sources.

3 From government's perspective:
– the assessment that certification requires appears to duplicate existing forest control and monitoring activities; and interferes with 'sovereignty';
– certification acts as a 'non-government check' on government activity;
– if certification is successful, the need for supervision by government is reduced.

For each of these perspectives, the solution framework is to establish an appropriate management system common to the individual forest owners. This management system would cover the LFMU to be certified. Such an option can be viewed from each of the perspectives outlined above:

Small woodland owner context and action

Certification appears to compromise wood production in favour of conservation. Yet planning for multiple forest use and structure can satisfy quality forestry objectives and maintain long-term levels of wood production. Multiple forest use requires more sympathetic management planning and more careful intervention.

A woodland owner certification initiative could be to:

- form an association – either use an existing association, or establish a new grouping as 'quality forestry participants';
- produce preliminary policy statement on 'quality forestry' for local conditions;
- communicate policy details and promote understanding of implications;
- assess the extent to which the forestry activities of existing association members meet defined policy;
- communicate assessment results and produce programme for further development;
- ensure consistency of documentation – especially management plan;
- ensure consistency of practices;
- develop an association information programme to inform, advise and produce targets for quality forestry;

- introduce internal control and compliance programme to ensure adherence;
- produce a register of association members (assistance in marketing and promotion of member's wood).

The advantages of this are:

- woodland owners can promote sales through identification as suppliers of certified wood;
- certification costs are reduced through the cooperative approach;
- programme of change is introduced to promote improvements in forest management;
- the system can be assessed and sources certified using field samples (which is cost-effective);
- association members may receive other benefits – marketing etc.

Industry context and action

Flexible wood procurement carries a supply risk, therefore stocks are held to compensate. Yet organized wood procurement reduces the supply risk, and stocks can be reduced. Organized wood procurement requires more efficient management systems and enables companies to become 'inspectable' for certification.

An industry certification initiative could:

- assess existing wood procurement systems – minimum requirement is that these identify forest owners;
- produce wood procurement policy to inform woodland owners of quality forestry objectives;
- assess, by sampling, forestry practices of wood sources;
- communicate assessment results, and act on areas where improvement is required;
- organize harvesting and wood delivery schedule to fit precise needs of the mill;
- develop supplier extension programme to inform, advise and produce targets for quality forestry;
- examine forest management plans of woodland owners to assess if quality forestry criteria are being addressed;
- produce register of 'approved suppliers' and eliminate poor suppliers after an adjustment period.

The advantages of this are:

- stocks are held on the stump; risk of wood deterioration is reduced; stocks are 'growing'; and woodland owners are identified by the system;

- a programme of change is introduced to promote improvements in forest management;
- the system can be assessed and sources certified using field samples;
- industry gains a greater understanding of woodland owner problems and future supply potential.

Government's context and action

Government and the national forest service already have in place and on the ground, using professionally qualified personnel, a system of extension, supervision and regulation. But the existing system may not be working to levels of 'quality forestry' now expected by society. Government systems and national forest service systems can be used as components of certification – although they may not provide all the elements required. For this to happen: (a) the extent to which existing systems cover generally acceptable quality forestry practices should be set out; and (b) independent assessment should be allowed for.

A government certification initiative could be to:

- assess existing coverage of system – minimum requirement is that forest owners and management plans can be identified;
- produce revised documentation where required – involve stakeholders;
- communicate activities and introduce appropriate systems (phasing in regions);
- ensure that desirable quality forestry practices are communicated, documented and implemented;
- assess forestry practices on the ground;
- communicate assessment results and commission credible independent assessment of activities.

The advantages of this are:

- as supervision mechanisms and forest practices are independently assessed, a forest area can be promoted as 'quality forestry';
- industry is supplied with the environmental specification required for raw material purchases;
- government (national forest service) receives an annual assessment of activity, against which performance can be measured and improvements planned;
- the national consumer and tax payer is provided with independent information on forest service performance;
- all forests in a particular geographic area are considered, reducing need for 'chain of custody' inspections.

In mixed rural areas, certification may also have to adopt a more integrated *land*

Box 8.2 *The move to integrated and balanced land use in the UK: LEAF*

LEAF: Linking Environment and Farming – an integrated crop management [ICM] project

LEAF was established in 1993 to develop and promote systems of agriculture which meet economic, environmental and agronomic criteria.

LEAF has two main objectives:
a) *Development* through ICM, as a combination of responsible farming practices which combine care and concern for the environment with the responsible and economic use of modern farming methods.
b) *Promotion* through selected LEAF farms throughout the UK.

LEAF is organized with:
a) an *Advisory Board* with members from the food industry, food retailers, ADAS [Agricultural Development Advisory Service], NFU [National Farmers Union], conservation groups, public interest groups, consumer associations, farming and the crop protection industry;
b) an *Executive Committee* which answers to the Advisory Board; and
c) a *LEAF Project Coordinator*

LEAF has produced guidelines for participants, with an assessment programme to ensure compliance. The guidelines require:
a) Commitment to the LEAF concept, objectives and regular assessment.
b) Understanding and adoption of the LEAF whole-farm management system. To suit farming to the site; to maintain or improve soil fertility on cropped land; and to achieve agricultural biodiversity.
c) Farm practices which ensure:
    •   adequate communication and training
    •   use of crops and varieties which minimize pest and disease incidence
    •   timeliness of operations so as to minimize environmental damage
    •   use of modern techniques aimed at optimizing use of farm chemicals
    •   careful management and use of crop residues
    •   maintenance of comprehensive farm records
    •   underlying ICM principles include post-harvest treatments
    •   all legal requirements are satisfied.

The assessment programme aims to promote and assist with improvement. It assesses:
a) landscape features;
b) wildlife habitats;
c) management of the soil;
d) crop protection;
e) conservation of energy;
f) pollution control; and
g) organization and planning.

In other words, its assessment is far broader than agronomic aspects alone. Forest certification programmes may consider adopting a similarly broad approach.

Certification in Practice

use approach. Woodlands are often part of the landscape and forestry an integral part of the farming system. Forests often occur in small blocks and strips that provide important wildlife habitats and amenity value. In these areas, common in much of Europe, single-issue certification programmes – like those currently being promoted – may not be the most appropriate. More integrated schemes, which take into account the whole farm system or landscape may provide a better result, both in terms of sustainable land use and in meeting the needs of the owner. Details of one such integrated scheme, the UK LEAF programme, are given in Box 8.2.

Application

Following market research or direct requests from buyers a forestry operation decides to consider the option of certification. Following discussions and an understanding of the processes involved in certification the forest manager makes a formal application to a certification body.

The application includes a preliminary evaluation often accompanied by an interview, discussion and presentation of what is involved. The certification body requests the forestry operation to submit copies of a forest policy or statement specifying the operation's environmental objectives. The certification body will also request preliminary documentation which demonstrates that the environmental issues in the policy are being addressed. If a management plan has already been prepared a copy of this will also be required by the certification body.

This initial review of the operation's environmental and forestry policy allows the certification body to evaluate if the company is ready to enter the certification programme immediately or if further development of management practices is required. Such a process avoids the operation using scarce funds, by going through the certification process prior to being in a position to potentially receive a certificate. If the certification body deems the LFMU's forest management practices are likely to meet the requirements of the certification programme, the certification process is put into motion.

During application the certification body formally undertakes to maintain complete confidentiality with regard to the LFMU; and provides formal details of the conditions attached to the certification programme. These will generally include:

- scope of the certification programme operated by the certification body;
- legal status and organization of the certification body;
- general conditions for obtaining and retaining a certificate – such as the provision of relevant information and acceptance that remedial action may be required between assessment and award of the certificate;
- requirement to appoint a designated staff member at the LFMU to liaise with the certification body;

- property and validity of the certificate – normally ownership of the certificate is retained by the certification body;
- right of access for surveillance visits by the certification body;
- notifications – such as for reassessment or for material changes in the LFMU management system made during currency of the certificate;
- publicity of award of the certificate and the marking of products from certified forests;
- circumstances under which suspension, withdrawal and cancellation of the certificate would occur; and
- appeals and complaints procedures.

Scoping visit

Once an application is accepted, the certification body may visit the forest in question. This is undertaken by a lead assessor with the objective of finalizing the assessment methodology; ensuring that the selected team has skills appropriate for the particular site; and finalizing the plan of work with field management.

Selection of the assessment team must ensure that adequate professional skills are available to address the priority environmental and social effects of the forestry operation.

A scoping visit is often required and its importance should not be under estimated. The scoping visit is often the first time that the certification body comes into contact with the LFMU. It follows that the scoping visit is often the first time that site management meet with personnel from the certification body. An important part of the scoping visit is detailed meetings with senior field staff to ensure complete understanding of the certification process and to ensure that the assessment logistics are feasible in the time and with the resources allocated. During these discussions the lead assessor will conduct a preliminary review of management documentation so as to become familiar with the particular operation 'style' and 'culture' as well as to ensure that coverage is adequate. The lead assessor will also gain a brief overview of the company, its departments, structure and geographic distribution of the forest area.

Specifically, a scoping visit would aim to address:

- an introduction to the certification body and presentation of the certification process and context;
- confirmation of the scope of certification required;
- explanation of the assessment – including both document review and field assessment – and the need for openness;
- nomination of, and agreement on, the member of staff from the LFMU who is to accompany the assessors during their work;
- explanation that, during the assessment, evaluation is undertaken by sample and problems may exist which are not detected in the initial assessment;

- confirmation of confidentiality;
- explanation of major and minor requests for corrective action; and that the raising of these does not necessarily mean a re-assessment; and
- fixing of dates for the assessment to start.

Document review [examination]

Figure 8.3 illustrates the basic requirements of an assessment. It starts by looking at documented management systems. The forest manager supplies all necessary documentation to the certification body which maintains a log of those submitted. The assessment will start only following a successful scoping visit; or if discussions with the forest manager indicate the LFMU is ready. In particular, the scoping visit report may recommend that identified deficiencies are corrected prior to the assessment. Once notification has been received from the forest owner that all aspects of the scoping visit report have been addressed the assessment can start.

Where a scoping visit has taken place, many of the items listed in the previous section would be addressed in an opening meeting with LFMU management. In addition, the assessment team would – during the opening meeting – record attendance and arrange a date and venue for a formal closing meeting. Following the opening meeting, a familiarization tour is made of the premises prior to commencement of the formal assessment.

In some instances, the forest operation will have supplied copies of key documents to the certification body prior to the assessment. This often takes place during the scoping visit. Under such circumstances evaluation of relevant documentation can start prior to the assessment. Such documentation could include final versions of the environmental policy, forest management plan, and/or operating procedures. Where documentation can be provided prior to the assessment the time allowed for reviewing documents on site can be shorter; and is usually more productive, as the assessor will have had more time to consider the submitted documentation and to consult over particular points before arriving on site.

The list of requested documentation will have been included and agreed to earlier – either in the scoping visit report or in written correspondence from the certification body. The certification body evaluates the LFMU's documentation for compliance with the certification standards. Key environmental and social effects must have been identified and prioritized; management systems should be clearly described including objectives and targets; and an assessment made of resource requirements. In particular, the audit team will check:

- the adequacy of the forest organization's environmental policy;
- that documentation satisfies national regulatory requirements;
- that social elements have been accounted for;
- that the forest operation allows for optimum use of extracted forest

resources and reduced waste from external resources used;

- that the environmental impact of forest operations is correctly addressed; and
- that forest management systems are robust enough to realize the objectives and targets set by the management.

To assist in the work, the assessor will use a document questionnaire provided by the certification body. The document questionnaire is signed by the assessor who completes the evaluation and any omissions or non-compliancies detected are listed along with any other queries. These are addressed to the forest manager in writing as soon as possible.

In exceptional circumstances, the amount of corrective work required may be substantial. The lead assessor may then recommend that no further assessment work is undertaken until the LFMU has taken the necessary corrective action and resubmitted the relevant documentation.

Field assessment [examination/validation]

The second part of the assessment [see Figure 8.3] involves an examination of internal and external site indicators and a validation of the documented management system. Internal site indicators would include ongoing research programmes, permanent sample plots, key conservation sites, etc. External site indicators include interviews with external stakeholders directly affected by the LFMU's activities; they might also include downstream effects in important water catchment areas.

Validation of the management system includes a sample check of described procedures to ensure their adequate field implementation. As part of the document review the assessment team would produce a number of assessment checklists. These are based on the documentation reviewed, and not on the certification standards. The objective of the checklists is to permit a logical and structured evaluation of field implementation. The items to be checked should be referenced to the documentation concerned and communicated to site management. Each assessment checklist is signed by the assessor who completed it, and is countersigned by the lead assessor.

Communication of the checklists to site management enables an assessment itinerary to be prepared and agreed upon between the assessment team and the LFMU. The checklists tend to place most emphasis on inventory and other resource assessment results; harvesting and road building activities; treatment of watercourses; and incorporation of special conservation needs.

In all cases, assessors consult personnel working in the LFMU who are responsible for the procedures being evaluated. This is in order to ascertain the level of understanding of the procedures and management plan; and – most importantly –the level of adherence to the procedures. Objective facts, including

Certification in Practice

93

Policy document

As a minimum, should contain a commitment to:

1 comply with all legislation
2 fulfil responsibilities to local people
3 optimize forest use and minimize waste
4 understand and minimize environmental impacts of all activities
5 ensure long-term productivity of the forest resource

In addition the policy:

1 must be endorsed by top management
2 must be publicly available
3 must address generally accepted principles of quality forestry
4 must be effectively communicated to all employees
5 must be reviewed and updated as required

Regulatory framework

As a minimum the LFMU must demonstrate conformity to:

1 forestry and land use laws
2 regulations on premises, equipment and staff facilities
3 labour laws
4 health and safety regulations

This requires the LMFU to:

1 maintain a register of current legislation
2 have procedures for communicating legislation to relevant staff
3 have a system for auditing its own compliance with legislation
4 have procedures to address identified non-compliance
5 maintain details on all external regulatory authorities
6 maintain records of all internal and external controls

Social elements

As a minimum the LMFU must ensure that:

1 land and resource use rights of local people are respected
2 concerns of local people are sought and taken into account
3 local people are involved to the fullest extent

This requires the LMFU to:

1 identify all local people affected by its activities
2 appraise the impact of its activities on local people
3 incorporate the results of the appraisal in its planning
4 demonstrate its right of custody over the forest resource
5 take account of sites of specific importance to local people
6 satisfy all legal requirements on working condition standards
7 recognize worker rights to organize for voluntary negotiations
8 have a defined mechanism for resolving disputes
9 continuously consult with and provide equal opportunity for employment and training of local people

Optimizing benefits from the forest

As a minimum the LFMU must ensure that:

1 it has a programme for optimal use of forest goods and services
2 yield of forest goods and services is sustainable in the long term

This requires the LMFU to:

1 maintain the widest range of forest goods and services
2 encourage the use of non-traditional wood species
3 encourage the development of downstream processing
4 investigate and implement waste reduction programmes
5 endeavour to use by-products
6 permit access to the forest for recreation and collection of NTFPs
7 gather data on sustainable production levels of all forest resources
8 monitor exploitation rates of all forest resources

Forest management systems

As a minimum the LFMU must ensure that:

1 its defined policies, objectives, targets, and procedures are integrated into a management system which is monitored

This requires the LFMU to:

1 have a written management plan appropriate to its scale

2 ensure that all staff are adequately trained for the job

3 carry out regular monitoring of its activities and contractors' activities

4 implement a system to identify products leaving the forest

5 ensure that objectives and targets embody a commitment to continual improvement

6 exclude high risk areas from logging and road construction

7 provide adequate buffers around significant cultural and biologically sensitive areas

Environmental impact of forest management operations

As a minimum the LFMU must ensure that:

1 the negative effects of its activities upon biodiversity of flora and fauna, landscape, soil condition and quality and quantity of water resources are minimized

This requires the LFMU to:

1 define and document responsibilities for environmental issues

2 define environmental issues in agreements with contractors

3 have assessed the environmental effects of all its activities

4 have operating procedures which address environmental effects

5 demonstrate that operating procedures are implemented

6 establish environmental objectives and targets

7 not use WHO 1A or 1B pesticides or chlorinated hydrocarbons

Figure 8.3 *Basic assessment requirements for certification*

Notes:
1 Documents should be appropriate to the scale of operations. A small-scale community forest would be expected to have fewer and simpler documents compared to a large industrial operation. For example, a policy document may consist of a single sheet of paper or it may be the first chapter in a procedures manual.

Box 8.3 *Corrective action requests [CARs] – a means of continual improvement*

Corrective action requests [CARs] are a common tool in the maintenance of ISO 9000 certificates. They can also be used effectively in forest certification. CARs are the primary way in which certification can produce change over time. CARs are the means by which the certification body ensures that continual improvement actually takes place.

CARs can be either *minor* or *major.* A minor CAR is raised when a single observed lapse has been identified in a procedure required as part of the LFMU's management system. A major CAR is raised where there is an absence or a total breakdown of a procedure required as part of the assessed organization's management system. Where a non-compliance is likely to result in an immediate hazard to forest quality, then the CAR shall be categorized as major.

Minor CARs raised during an assessment or during surveillance do not preclude the LFMU from being certified. Once a minor CAR has been raised, the organization shall respond in writing within three months detailing the actions taken to prevent recurrence of the problem. The effectiveness of the resulting action taken by the LFMU must be verified at the next surveillance visit. At that visit, the assessor shall decide whether or not appropriate corrective action has been taken. The action taken must have been implemented for a sufficient period of time to allow adequate evidence to be available to the assessor when making this decision.

Major CARs raised during an assessment or during surveillance preclude the LFMU from being certified. Following a major CAR raised during assessment, the certification body should be notified of the action taken within one month of the CAR issue date. The action, as notified by the LFMU, shall be verified by the assessor within two months of the CAR being issued. A major CAR raised during surveillance is regarded very seriously. The certification body is notified of the action taken by the LFMU within two weeks of the CAR being issued. This action is notified by an assessor within one month of issue, regardless of whether notification of action has been received or not.

When CARs are raised they are written up following a specific format and are signed by both the assessor raising the CAR and the LFMU. When a follow-up visit shows that corrective actions have been satisfactorily implemented, the assessor signs and dates the CAR as acceptance of the corrective action taken.

records and site evidence, are examined to substantiate the adequacy of compliance both with the LFMU documentation and the certification standards. At any time, the assessment team may consider aborting the assessment due to a high level of non-compliance in evidence. This decision is made by the lead assessor in consultation with other members of the assessment team; and is based on both the degree and amount of non-compliance. Should the LFMU request that the assessment process continue, then this is acceptable provided that the LFMU agrees that the current assessment is technically aborted and that a complete re-assessment would take place at a later date.

At the end of the field assessment the assessment team assembles to determine compliance with the certification standards and to prepare a draft assessment report. The assessment report should demonstrate:

- that an assessment has been undertaken;
- the manner in which the assessment process was conducted;
- the results and conclusions of the assessment process; and
- the decision as to whether or not to recommend that award of a certificate be given.

The initial assessment results are presented to, and discussed with, the LFMU management prior to the assessment team's departure at a closing meeting.

The closing meeting would typically address:

- presentation of findings and reporting on decisions;
- explanation of decisions regarding major and minor actions requiring correction;
- obtaining a signature from an authorized representative of the LFMU to all agreed actions requiring correction;
- obtaining a signed confirmation from the LFMU that the assessment has taken place;
- explanation of the peer review process; and
- recording any disagreements with findings.

Following the closing meeting, a package of assessment documents is produced for presentation to the LFMU and submitted to peer review. The package of documents would contain the assessment report; requests for corrective action [see Box 8.3]; copies of assessment checklists, assessment itinerary, documentation questionnaire, scoping visit report and pertinent correspondence.

Peer review [validation]

The assessment report and associated documentation, as required, is sent for peer review by at least three independent specialists. These will have been selected for their experience and knowledge of the forest type in question; technical expertise; and international standing.

The primary function of the peer review process is to attest to the technical credibility of the assessment methodology of a particular certification exercise and to examine the conclusions reached by the assessment team. The peer review process is, therefore, critical in adding a second tier of professional expertise to the assessment prior to the decision being taken as to whether a certificate can or cannot be awarded.

The role of the peer review is to ensure that the assessment report has the necessary content to act as the foundation for the award of a certificate and to confirm that the assessment team has:

Certification in Practice

- carried out an objective and professional assessment;
- investigated all relevant data sources and avenues of enquiry;
- arrived at an appropriate conclusion based on the evidence presented to it; and
- prepared a concise and quality report that will stand up to public scrutiny.

The peer review process underwrites the quality of the assessor's work and assists in providing the assessment decision with the support that will give the certificate international credibility. Individuals to be included as peer reviewers should be approved by the governing board of the certification body. In order to maintain quality and consistency of the peer review process the certification body should define and document a set of procedures that cover the peer review scope.

Certification

Following approval of the assessment recommendations by the peer review process the LFMU may be awarded with a certificate. Award of the certificate is accompanied by responsibility for its maintenance. This requires a commitment to continual improvement of environmental and social performance; and an undertaking to fulfil any requirements for immediate corrective action which have been recommended.

The certificate remains the property of the certification body and should not be copied or reproduced in any manner without the prior approval of the certification body. Any modification to the forest management practices or forest area of the LFMU should be reported to the certification body. The certification body will determine whether or not the notified changes require additional assessment. Failure to notify the certification body can result in suspension of the certificate.

The LFMU has the right to publish that the forest in question has been certified and to apply the certificate mark to stationery and promotional material. In so doing the LFMU should ensure that no confusion arises between certified and non-certified forest areas. The LFMU should not make any claim that could mislead purchasers to believe that a product derives from a certified forest when, in fact, it does not.

The certification body can suspend the certificate for a limited period where corrective action requests have not been signed off in the time agreed; or where incorrect or misleading references have been made in respect of the certificate. At the same time the certification body should indicate the conditions under which the certificate can be reinstated. If these conditions are not fulfilled the certificate should be withdrawn. At all times the LFMU has the right of appeal.

Notification of an LFMU's intention to appeal should be made in writing to the certification body – usually within a specified time limit of notice of certificate withdrawal. Appeals are judged by a sub-committee of the governing board comprised of at least three non-executive members. The certification body is required to submit evidence to support its decision. The decision of the sub-committee should be final and binding on both the LFMU and the certification body.

Labelling [chain of custody]

Where the forest manager and/or buyers of wood from the certified forest wish to identify the wood as coming from a certified source, it is necessary to apply for chain of custody inspections. Chain of custody can be defined as: 'an unbroken trail of acceptability that ensures the physical security of samples data and records.'[1] As with certification of the forest area, it is important to differentiate between a chain of custody 'system' – that which is installed by the various parties in the chain; and chain of custody 'assessment' – which relates to the activities of the certification body in order to provide a verification of product origin.

Figure 8.4 illustrates typical assessment requirements for chain of custody inspections. To varying degrees, chain of custody requires that products are identified and segregated; and are accompanied by a system of records which can be easily interpreted. The chain of custody must be able to provide physical evidence that the certified product originates from a particular source; requiring a secure data capture and communications system which runs in parallel with and links to the physical evidence.

To some extent, there is a trade-off between the need to identify and the need to segregate certified products. An efficient and easily-recognizable identification and recording system may reduce the need for segregation. In all cases, the application of chain of custody systems should use techniques and technology which are appropriate to the product. For example, the transport and manufacture of high-value wood products from large logs can justify a sophisticated product identification and recording system related to individual pieces. In contrast, composite wood products using low-quality material – often in particle form – will require a system whose emphasis is on product segregation and batch identification.

Chain of custody is a critical element of any certification programme since it provides the link between buyers and sellers from the forest to the point of final sale. Box 8.4 describes the rationale for chain of custody inspections in more detail. It is important, for credibility to be maintained, that the chain of custody remains *intact* throughout, particularly at stages where responsibility for the goods changes. Essentially, chain of custody is a stock control exercise which requires the goods to be secure and requires transparency for ease of inspection.

The chain itself will consist of a number of links; the number depending on the range of sources, the complexity of the manufacturing process and the type of market into which the product is sold. An example which shows the chain of

---

1 Quality Systems Requirements for Environmental Programs, ANSI/ASQC E4-1993, May 1993. American Society for Quality Control, Energy and Environmental Quality Division, Environmental Issues Group.

Principal criteria

An organization wishing to identify products from certified forests must:

1 provide physical evidence that the goods originate from a particular forest
2 identify and segregate the goods concerned
3 document the record system which runs in parallel to the goods
4 monitor compliance with chain of custody requirements

Product identification

As a minimum the organization must ensure that:

1 all products from certified forests, or manufactured from products derived from such forests, are clearly marked as such
2 documented procedures exist to control the marking of certified products

From the forest to the mill the organization must:

1 mark logs in order to identify the forest of origin
2 record the volume of logs transported from the forest by species

From primary conversion to subsequent stages of manufacture, wholesale distribution and retail the organization must:

1 mark products to identify the production run or stage of repacking
2 record the volume and species or type of product at each step

Product segregation

As a minimum the organization must ensure that:

1 all products from certified forests, or manufactured from products derived from such forests, are segregated from other products
2 documented procedures exist to control the segregation of certified products

From the forest to the mill the organization must:

1 segregate certified logs from non-certified logs in the log yard
2 implement a batch production process to segregate certified products during processing unless automated coding mechanisms are used
3 segregate certified products in the production area, green timber store, kilns and final storage areas

From primary conversion to subsequent stages of manufacture, wholesale distribution and retail the organization must:

1 segregate all certified products upon arrival at the premises
2 implement a batch production process
3 prepare separate documents for certified products during storage and shipment

Records

As a minimum the organization must ensure that:

1 orderly records are kept relating to purchase, shipment, receipt, forwarding and invoicing for certified forests
2 records include customs, phytosanitary, transportation and invoice documentation

The organization must:

1 monitor its record keeping system
2 show that transport and invoice documentation can be reconciled with actual loads
3 ensure that documentation is transmitted ahead of certified products

Figure 8.4 *Basic chain of custody requirements*

Box 8.4 *The rationale for chain of custody inspections*

Tracking of forest produce
A certificate of quality forest management is awarded to those who can demonstrate that their forests are managed well, and who show genuine long-term commitment to improving forest management. Yet physical distances between producers and consumers are great; and the wood may pass through many hands on its way from the forest to the final point of sale. Consumers need reassurance that wood-based products they receive are genuinely from the forests specified. Chain of custody inspections aim to provide consumers with this guarantee.

What is the relevance of a certificate covering the chain of custody?
Many end-use markets – including retailers, municipalities and specifiers – demand independent proof of the origin of forest goods. A certificate of quality forest management goes half-way to meeting market demands. Chain of custody inspections are required to provide complete confidence in the forest produce when it arrives at the final point of sale.

•   Chain of custody inspections underpin a certificate of quality forest management, where products are identified as originating from such forests.

What advantage will chain of custody inspections give?
Exporters of wood produce will be able to identify the origin of their raw material and in turn pass this information on their customers. In addition, the process of installing inspectable chain of custody systems can bring a variety of direct benefits:

•   improved stock control and business efficiency; and
•   increased understanding of consumer markets by enabling more direct communication between ends of the supply chain.

---

custody requirements for the manufacture and export of doors from rubberwood is given in Figure 8.5.

Each organization in the chain should establish and maintain procedures appropriate to its scale for identifying individual products or batches from particular sources. Each identification should be unique and recorded. Through the identification and associated records, it should be possible to trace the product to its immediate source; original shipment and/or batch; and certified source of origin. It should also be possible to complete an input:output audit at each organization in the chain. The quantity of certified material bought by the organization should approximate to the amount sold after allowing for processing losses. Usually an appropriate conversion factor and acceptable tolerance limit are agreed upon between the organization in the chain and the certification body.

Where appropriate, each organization in the chain should allocate a new identification at the time of receipt of goods. Where a batch production process is

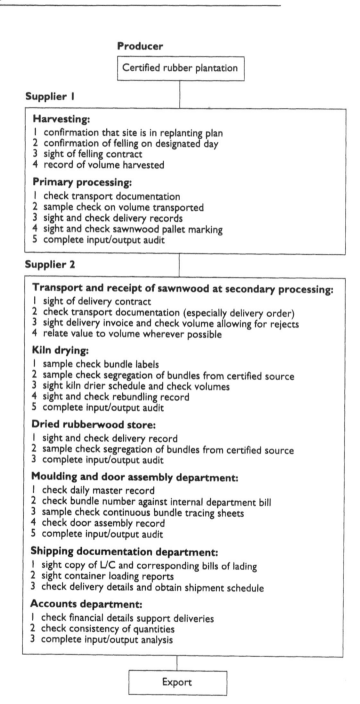

Producer

Certified rubber plantation

Supplier 1

Harvesting:
1 confirmation that site is in replanting plan
2 confirmation of felling on designated day
3 sight of felling contract
4 record of volume harvested

Primary processing:
1 check transport documentation
2 sample check on volume transported
3 sight and check delivery records
4 sight and check sawnwood pallet marking
5 complete input/output audit

Supplier 2

Transport and receipt of sawnwood at secondary processing:
1 sight of delivery contract
2 check transport documentation (especially delivery order)
3 sight delivery invoice and check volume allowing for rejects
4 relate value to volume wherever possible

Kiln drying:
1 sample check bundle labels
2 sample check segregation of bundles from certified source
3 sight kiln drier schedule and check volumes
4 sight and check rebundling record
5 complete input/output audit

Dried rubberwood store:
1 sight and check delivery record
2 sample check segregation of bundles from certified source
3 complete input/output audit

Moulding and door assembly department:
1 check daily master record
2 check bundle number against internal department bill
3 sample check continuous bundle tracing sheets
4 check door assembly record
5 complete input/output audit

Shipping documentation department:
1 sight copy of L/C and corresponding bills of lading
2 sight container loading reports
3 check delivery details and obtain shipment schedule

Accounts department:
1 check financial details support deliveries
2 check consistency of quantities
3 complete input/output analysis

Export

Note:
At every stage of the process: (1) output must stay within a reasonably consistent % of input; (2) rejects must be duly recorded; and (3) output must carry a sequential serial number which can be related back to inputs of fresh wood.

Figure 8.5 *Identification of chain of custody with regard to the manufacture of doors from rubberwood*

Certification in Practice

used, it may be more appropriate to allocate a new identification at the end of the production run and to the packed bundle. Under such circumstances it should be possible to trace the product to a particular production run; and hence through the associated documentation to the various raw materials used in the process.

Certified product ideally should be stored separately from non-certified product. Documented procedures appropriate to the scale of the organization should exist to ensure that non-certified product is prevented from inadvertently entering the production process.

By now it should be clear that good records are a key element for successful chain of custody assessment. All records must be legible and easily identifiable to the product involved. Each organization in the chain should aim to maintain the following records:

1 **Purchase records [not relevant for LFMUs]:**
   - purchase orders, contracts, invoices and list of suppliers of certified product;
   - quantity, nature, species and origin of certified product;
   - quantity, nature, species and origin of non-certified product; and
   - goods inwards notes and records of delivery acceptance and inspection.
2 **Stock records:**
   - stock records of raw materials and finished product, including – where appropriate – periodic stock-taking results;
3 **Production records:**
   - job cards, production orders and other milling and processing records;
   - quantity, nature, species and origin of certified raw materials used;
   - quantity, nature, species and origin of non-certified raw materials used;
   - details of all product manufactured and quantities produced;
   - batch numbers and records; and
   - wherever possible agreed conversion ratios should be assessed in order to confirm their accuracy and that no non-certified material is entering the production process.
4 **Sales records:**
   - sales orders received;
   - details of products and quantities sold, paying particular attention to descriptions provided both by and to the customer; and
   - batch details where appropriate.

Publicity material used, and claims made by the organization concerning the source of origin of the product sold, would also be verified as part of the assessment.

Initial meetings between the organization in the chain and the certification body would agree upon acceptable procedures for product identification, segregation and record keeping. The cost of the assessments can be reduced if the

Certification in Practice

organization implements a structured programme of its own for internal audits of the agreed system. Such audits should be planned in advance and document-ed. They should also aim to verify that the activities carried out within the organization comply with the planned and documented arrangements described; and measure the effectiveness of them to meet the declared objectives. In com-plex situations, a programme of internal audits would be essential.

Periodic review [surveillance]

The extent of periodic review required is determined by the assessment report; and particularly by the number and degree of corrective action requests. The assessment report will set out the initial timetable of surveillance visits required as well as the particular aspects of the LFMU's activities that require attention.

The assessor who is to undertake the surveillance visit would obtain the pre-vious surveillance report (or the assessment report if it is the first surveillance visit); details of corrective action requests; and any complaints or appeals which have gone on file since the last surveillance visit.

The assessor should contact the client in order to arrange for a date. A six-monthly frequency of visits should generally be maintained by the certification body; with visits permitted to take place two months either side of the nominal date. A surveillance visit should:

- cover at least 20 per cent of the LFMU's management system; and in particular should address any changes which have been made since the previous visit. The assessor should also try to cover areas of the management system which were not addressed in previous visits;
- verify that all observations made during the original assessment have been acted upon;
- verify that all due minor Corrective Action Requests have been dealt with;
- audit the LFMU's procedures for internal monitoring;
- aim to consult LFMU personnel who are responsible for the procedures being assessed;
- validate the effectiveness with which the management system assessed is being implemented;
- examine promotional materials to check that there is no misrepresentation of the certificate; and
- raise non-compliances in the form of Corrective Action Requests where appropriate.

On completion of the surveillance visit, a report should be produced which is signed by the assessor and a representative from the LFMU. Surveillance visits may also include chain of custody inspections and/or checks to determine whether required changes to the chain of custody system have been made.

Cost

Certification has two types of cost: the cost of certification itself [direct costs] and the cost of restructuring required to achieve certification [the indirect costs of achieving quality forestry]. In many instances, the latter is likely to be more difficult and expensive than the former. It follows that the cost of a wholesale transition to systems of quality forestry is likely to be more than the direct costs of certification.

Direct costs

These are heavily influenced by the feasibility and ease of assessment by the certification body. This, in turn, is influenced by the extent to which the LFMU has strong and transparent management systems in place. Strong management systems reduce the amount of field assessment and validation required. Insistence on strong systems means that the costs of certification are kept down. In addition, if LFMUs are encouraged to install quality systems these may provide other benefits.

For many forestry activities good environmental management is good business. For example, harvesting activities which cause erosion can also result in reduced site productivity, lost long-term income, sedimentation of watercourses and criticism from downstream water users.

Some certification costs are site-specific and depend upon the size and complexity of the LFMU, extent of biological variability within the forest, diversity of the social environment, and the degree to which activities are documented in clear systems. Experience to date with certification does, however, allow some general pointers to be made.

Based on experience so far, the minimum cost for an assessment tends to be about US$ 500. This should be sufficient to assess a small farm woodland close to the assessor's head office. For natural forest, to this should be added an average of about US$ 0.40 per ha for the initial assessment and US$ 0.15 per ha for subsequent surveillance visits and chain of custody assessment. Plantations with faster growth would, proportionately, cost slightly more per ha – although the increase would usually be offset by greater revenue from such areas. There is not currently enough experience with the certification of plantations to provide more specific figures. It should be pointed out that, in all cases, the extent and costs of surveillance requirements depend considerably on the outcome of the initial assessment. For a certificate valid for six years, a 100 000 ha natural forest concession in the tropics would incur total costs of about US$ 130 000 over the life of the certificate [ie about US$ 22 000 per year]. One hundred thousand hectares is the size of Greater London.

Certification in Practice

Indirect costs

These are associated with installing the systems which permit quality forestry. Experience has shown that investment in better planning, documented procedures, training, forest management and silviculture can yield positive long-term returns. Installing these systems requires increased planning and up-front investment; it also can result in a changed relationship between fixed and variable costs. Fixed costs per unit of output – especially roads – tend to increase. The only way to keep total costs down is to reduce variable costs per unit of output. This means improving productivity. The level of output at which this trade-off lies will vary from forest to forest, but it often means that average costs at low levels of output tend to be higher with certification than without. That is, certification of small LFMUs is likely to be proportionately more expensive than for large LFMUs. In certification there are economies of scale – hence the value in small woodland owners forming an association, as discussed earlier in this chapter. The changing relationship between fixed and variable costs is illustrated

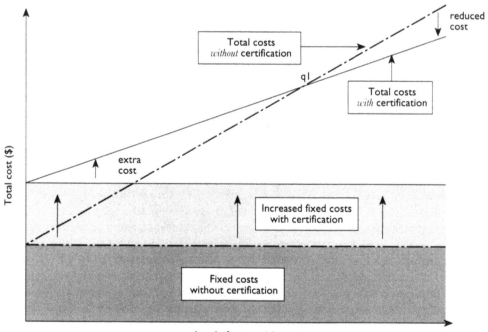

Notes:
1 Typically certification requires an increase in the ratio of fixed to variable costs.
2 Increased fixed costs require a reduction in unit variable costs in order to contain total costs.
3 q1 = level of output where total cost with certification equals total costs without certification. At higher levels of output the total cost per unit of production is lower with certification. At lower levels of production the total cost is higher with certification.

Figure 8.6 *Indirect costs of certification*

schematically in Figure 8.6. It should be noted that even where indirect costs increase, the effect on final product prices will be lower.

Case studies

To illustrate the effect of certification on actual forest operations, four case studies have been included in Boxes 8.5–8.8. Two of these are industrial-scale operations: Collins Pine in the USA and Demerara Timber Limited [DTL] in Guyana. Dartington Woodlands in the UK has been included as an example of a small and scattered forest operation; and Bainings in Papua New Guinea as a community forest project.

The motivation for certification was slightly different in each case, but all expressed the objective that certification would *improve sales* – either through higher prices or through better market access.

Box 8.5 *Case study: Certification of Collins Pine, USA*

Description:
The Collins Almanor Forest [CAF] is a 94 000 acre [38 000 ha] forest of mixed conifers owned jointly by the Collins family and the World Division of the United Methodist Church. The Collins Pine Company also operates a sawmill located in Chester, California, close to the CAF. CAF was certified in March 1993 by Scientific Certification Systems [SCS, USA].

Eighty per cent of CAF is of California mixed conifer type forest. This is a complex association of ponderosa pine, sugar pine, douglas fir, white fir and incense cedar, intermixed as single trees or in a mosaic of small single-species stands. This forest offers a rich range of wildlife habitats and is centred between 4200 and 5500 feet [1300 and 1700 metres] elevation. The remainder of CAF consists of the red/white fir type at higher elevations and ponderosa pine with black oak at lower elevations. Hardwoods include alder, cottonwood, maple, willow, dogwood and quaking aspen.

Forest management:
Management of CAF has been stable. Since 1941 there have been only four managers. Early management philosophy was to plan 'so that, in the minimum possible time, the area will be placed upon a sound, economic sustained yield basis, through sound forestry practices. This entails the use of selective logging, protection from fire and insects, plus maximum utilization and care for the reserve stands of timber, to assure a continuous supply of logs and other forest products. This continuous flow of forest wealth will provide for a high standard of living and stabilized community which, in turn, will benefit the community, County, State and nation as a whole'. Today, management and determination of the annual allowable cut is also driven by the biological capability and environmental constraints of the forest and is currently about 35 million board feet [140 000 m^3].

During the certification process, Collins learned that they had to modify some of

---

their forest management practices, including the incorporation of more wildlife trees and increased thinning of white fir to get back to a natural mix of species. Collins' investment in forest management has increased because of certification – they are installing GIS systems to give them better inventory information.

What motivated Collins Pine to opt for certification?

Marketing was the original reason for pursuing certification. The company had always had high stewardship principles in its forest management, and senior management saw an opportunity to reach the 'high ground' in the 'timber wars' occurring in the Pacific Northwest. The owners were quickly convinced that certification was the proper thing to do and that management of the CAF could withstand the outside scrutiny of an independent third-party.

The expected benefits were to: gain market advantage; validate forest management practices; assure shareholders that land is being managed sustainably; educate and raise the awareness of consumers and the public; position the company as an industry leader; improve the company's governmental and political clout; increase the company's credibility with environmental groups; raise the awareness and morale of the company's employees; and pursue certification as the 'right thing' to do.

Problems in certification encountered by Collins Pine and SCS:

No major problems were encountered with the Timber Resource Sustainability or Forest Ecosystem Maintenance elements of SCS's certification programme. Collins did have problems with the Financial Viability elements, because they are an extremely private firm which did not feel comfortable divulging financial information to outside parties. SCS was ultimately able to reach a mutually acceptable framework for the transfer of the necessary financial data.

The most difficult issue has been chain-of-custody. Collins provides about half of its own resource base at the Chester mill, and so is required to segregate certified timber from non-certified timber. Chain-of-custody requirements also present a challenge to some of Collins' customers, who make windows using components coming from a variety of different suppliers.

Collins would also like to see more of a commitment – in terms of purchases – from those within the industry who say they are advocates of certification.

Realised benefits of certification:

Collins believes that it made progress in gaining all the expected benefits listed above. Some *additional* points are:

- Certification helped Collins' marketing programme. Certification altered the product mix from principally commodity products to more value-added products.
- Certification helped Collins Pine in procuring outside timber from other private landowners (to supplement the timber from CAF to the Chester mill) because those landowners preferred to deal with an environmentally-sensitive company.
- Certification has improved company morale. Employees know that they will be at Collins for the long-term. This confidence is rare in an industry that is currently suffering from severe employment cutbacks.
- Collins has received a high degree of media attention as a result of certification. This positive attention has helped their public image.

What was the principal outcome of certification?

Commodity items, such as 2 x 4 inch boards, sold for no premium. However, of the potential certified wood – about half of Collins' total sawnwood production – Collins sell 15 per cent by volume into certified markets at an average of 11 per cent premium over non-certified products. This certified product goes into Do-It-Yourself markets, furniture and framing materials in communities with high environmental concerns. Collins Pine has acquired a contract with Lexington Furniture – a division of the largest furniture manufacturer in the United States – to provide them with a large volume of White Fir for a very high environmental-profile line of furniture. This project earns Collins tremendous attention within the furniture industry and the marketplace. The utilization of White Fir in this project has also been an excellent opportunity for Collins to increase the economic value of this species.White Fir, a shade-tolerant species, has increased in CAF due to the repression of fire. Collins has had to thin the White Fir in order to maintain the natural balance of species in the forest. The Lexington project provides them with a means to utilize this species, assisting them in their silvicultural objectives.

The reaction of the staff of Collins Pine:

Senior management was enthusiastic, but initially middle management was wary of change. They were generally sceptical of having outsiders coming to assess their forestry practices. As industrial foresters, they were fearful of the environmental reputation of certain SCS team members. This attitude was changed, as they found the SCS team members to be highly professional and found consensus on most forestry principles. Today, Collins forestry staff are among the strongest supporters of the third party certification effort. The family ownership feels that certification is one of the best and most rewarding efforts ever pursued by Collins Pine.

The clients' reactions:

Clients were generally passive. Most could not – or would not – comply with the chain-of-custody requirements. They were not getting enough demand to change their method of operation. However, Home Depot – a large retailer – was proactive and top management was supportive. Even here, middle management initially resisted change and it took a year to get the programme on track. Today Collins' shelving is promoted in Home Depot stores throughout the San Francisco Bay Area. Collins has found support for their products in 'green areas' like Austin, Texas; Sun Valley, Idaho; and Santa Cruz, California.

Source: Debbie Hammel, Director for Forestry Programs, Scientific Certification Systems, Inc., letter to IIED, October 1994

Box 8.6 *Case study: Certification of Demerara Timbers Limited, Guyana*

Description:
Demerara Timbers Limited is a private company which has a 50-year concession over 1.3 million acres [0.5 million hectares] of tropical rainforest in Guyana, South America. The company was a former State-owned company, divested by the Government of Guyana in July 1991 in an attempt to attract foreign investment into the country. From the outset, the new management of DTL aimed to develop a unique system for the sustainable management of the rainforest: the Green Charter.

The Green Charter is a list of ten principles of wise forest management based on scientific research and international consultation. Using detailed inventory plans and selective logging techniques, the company harvests an average of four trees per acre [ten trees per hectare] every 20 years. The Charter also commits the company to developing 'added-value' products and the development of a responsible waste management programme. The Charter forms the headline to a full Forest Management Plan, approved by the Government of Guyana and the Guyana Forestry Commission.

After the Green Charter had been developed, the management of the company began a marketing campaign to build upon the former company's existing export business. For many years, Guyana had been known as one of the prime producers of high quality marine timbers, specifically Greenheart [*Ocotea spp*]. DTL sought to develop this market further and to stimulate interest in so-called 'lesser known' species such as Andiroba, Kabukalli, Mora and so on. The Guyanese forests have many species which are useful for a variety of applications, including furniture production.

Although the Charter has won international support, DTL has also sought additional independent verification of its approach. This verification consists of several elements:

1   Scientists from the Tropenbos Programme, an independent research organization based in the Netherlands, live and work in DTL's forest. They operate independently from DTL, and the results of their studies are published.
2   The company runs its own in-house Green Audit Programme, which monitors adherence to the Green Charter.
3   DTL has formed an International Advisory Council, which includes high-profile professionals.
4   In June 1994, DTL was awarded a certificate of Quality Forestry by the independent monitoring organization, SGS Forestry. This certificate also stated that DTL was adhering to the principles of the Green Charter.

In the course of developing the DTL approach to forestry, DTL's management studied the wealth of scientific and environmental material available, and attended relevant conferences and seminars. It became obvious that the world was focusing on certification as one of the few reliable methods of verifying sustainability. Since DTL attempted to position itself near the forefront of sustainable timber production, application for certification seemed the natural course of action.

Benefits of certification:

No direct benefits of certification have yet been in evidence. First of all, it is difficult to quantify such benefits, since DTL is the first large tropical forest operator to gain a certificate, and restrictions in the international trade of tropical timber have yet to make themselves felt. Studies in the Netherlands have shown that, while certification is unlikely to lead to higher prices in the short term, producers of certified products may gain easier access to high-value *niche* markets, which are environmentally sensitive. Such markets would include local authority tenders for major sea defence work, for example.

Since DTL already had a comprehensive management plan, it cannot be said that the process of certification of itself led to improved forest management practices. This is because DTL was already using exhaustive inventory methods and careful harvesting in order to meet the demands of the Green Charter. However, the process of obtaining certification has undoubtedly improved the company's stock control, for example. The discipline of instituting a 'chain of custody' system has resulted in improved documentation and tracking systems, leading to management having greater control over the business.

DTL's staff have been used to working to strict environmental guidelines and there appears to be a fair degree of understanding and sympathy with the Green Charter's aims. In this climate, certification was well received by staff.

There is no doubt that achievement of certification has resulted in increased attention from potential timber purchasers. Existing purchasers have expressed positive views about the certificate, although there is a fair degree of misunderstanding of the value of various types of certificate which might be available.

DTL is the first timber-producing company in Guyana to have sought certification (and one of the first worldwide). Other timber producers have expressed some interest in the process, but so far remain largely unconvinced of the value of the certificate, particularly for the markets currently being served, such as the *local* Guyanese market and the Caribbean. Decisions on timber purchases in Guyana, for instance, are heavily dependent upon price.

It is also quite common for purchasers and members of the public to ask such questions as 'who are SGS?', 'who are the FSC?', 'what do Friends of the Earth think about it?' and so on.

DTL has suffered to a certain extent through being at the leading edge of certification. It would have been much easier and more meaningful to publicize achievement of the certificate, had FSC already completed their accreditation process. However, it is understood that SGS will be accredited by FSC very shortly and it is hoped that positive benefits will then start to flow through the certified timber companies.

Source: Janet Croucher, Demerara Timbers Ltd; letter to IIED, October 1994

Box 8.7 *Case study: Certification of Dartington Woodlands, UK*

Description:

The woods on the Dartington Estate in Devon extend to 190 acres [77 ha]. They are owned by the Dartington Hall Trust and are managed by Silvanus. The main woodland block is 100 acres [40 ha] – North Wood – and the remainder is made up of much smaller blocks of between 4 and 10 acres [1.6 and 4.0 ha].

Since their purchase in 1920 the woods have been used to experiment with innovative and, at the time, untested forestry systems and practice, mainly using conifers.

The woods are largely coniferous, but with areas of broad-leaved trees and mixtures. Some of the legacy of the experimental days remains, with a very wide species mix and very small compartments and sub-compartments.

Management objectives:

Until 1966, the overall aim was to yield the greatest possible profit. From 1966 to 1990 the woods were managed on a contract basis with Silvanus taking over management from 1990 to the present day.

The aim of management was changed in 1990 to maximizing the ecological value of the woodlands and this combined with the history of leading the way led the estate to look at timber certification.

The certification process:

The application to the Soil Association was made in March 1994 and, after considerable deliberation and debate, was approved in July 1994. Inevitably, as the first woodland in the UK to go through the certification process, there were teething problems and much to-ing and fro-ing.

The direct cost to the estate of the certification was about GBP 700 [US$ 1000] but complying with the Responsible Forestry Standards hasn't resulted in a fundamental change in management. A few minor changes were required, but on the whole the woodlands were being managed to a standard acceptable to the scheme.

The benefits at this stage are largely unquantifiable but it is hoped that the certified timber and timber products will command a price premium. Judging by the interest we have had, a premium of between 5 per cent and 10 per cent looks likely.

Silvanus, as managing agents of Dartington Woodlands, have been very supportive of the certification process and indeed were involved in advising the Soil Association on the drafting of the Responsible Forestry Standards. The resultant publicity following certification has been enormously beneficial to Silvanus but, as a consequence, continues to make demands on staff time. We are encouraging other clients to go for certification and will redouble our efforts if a premium can be achieved for the certified timber.

Source: James Lonsdale, Silvanus, letter to IIED, October 1994

Box 8.8 *Case study: Certification of Bainings Community Forest Project, Papua New Guinea [PNG]*

Background:

The Bainings Community-based Ecoforestry project is operated under the overall supervision of Ulatawa Diwai, a division of Ulatawa Estates Pty Ltd in East New Britain, PNG.

Ulatawa Estates is the last remaining expatriate-owned and -run plantation on the Gazelle Peninsula of East New Britain, producing a wide range of vegetables and flowering plants. The area is largely owned by clans (communities) of the Bainings Tribe. During 1990, Max Henderson (a naturalized PNG citizen and plantation owner) established a sustainable forest management programme with the assistance of the Riet Clan. The project is managed by Max Henderson with the assistance of 25–30 Bainings people, who receive a standard daily wage, while the clan landowners receive statutory royalties for the logged trees. Profits from the operation are divided between Ulatawa Estates and a trust fund established to promote community development.

The forest:

At the time of the assessment, the Ecoforestry Project consisted of forest management and wood processing activities in three forest areas and associated villages; those of Riet, Arambum and Maranagi. The area of production forest is 12,500 ha. Although forest of a similar extent is available in hilly terrain, it is not considered for production at present due to access and transport difficulties. These are currently designated as protection or reserve forest.

Salvage forest does not exist within the production area. Land suitable for reforestation is available in the form of land lying fallow after shifting cultivation. This land is, however, kept in store by the landowners for future cultivation, operating over a twenty-year cycle. Reforestation with the natural long-rotation species favoured by the project is, therefore, not an option in such areas.

In the low-lying and easily accessible areas [production forest], two main forest types exist: Kamarere [*Eucalyptus deglupta*] and Taun [*Pommetia pinnata*]. Kamarere appears as natural even-aged stands in the lower valleys. Taun appears throughout the area, but achieves better form on higher ground.

Forest management objectives:

The Project, supported by the Pacific Heritage Foundation and B&Q plc [a UK based DIY retailer], aims to:

1 promote an increased awareness amongst all people of the wealth and diversity of natural heritage in the Pacific;
2 improve the welfare of the people consistent with improved conservation and managed economic programmes.

The Project's operational forest management objectives include the following:

1 The Bainings Project wants to show that the sustainable harvesting of tropical timber by the landowners themselves is possible, desirable and economically viable.

2   The villagers will be encouraged to use their forest as a sustainable resource and the trees which are cut will be converted into timber (sawnwood) by the landowners themselves.

3   The landowners will be encouraged to see their forest as a perpetual asset, and to maintain and harvest that asset in a manner through which they can gain constant income, provide employment in areas adjacent to villages and remain completely in charge of their own resource.

Forest management operations:
Trees are cut using a chainsaw and sawn on site by the landowners using portable sawmills. The sawn timber is carried out by hand and transported by truck to Ulatawa, where it is processed further.

Prior to harvesting, temporary road lines are established. A pre-harvest inventory determines those trees to be cut and those to be left, eg seed trees, over-mature trees and those to be cut in the next cycle. The sawmill site will typically last for 6–12 months and will be replanted after use. A post-harvest inventory identifies the standing resource, immature trees and seedling type and growth.

Export-grade timber is purchased from the landowners by Ulatawa – offcuts are converted into garden stakes and the butts of trees skimmed to produce flat slabs for export. Timber for local sale is treated and dressed while timber for export is dried in solar kilns. Currently, this export grade material is being imported into the UK, where it is further processed and purchased by B&Q.

Forest assessment:
Certification should reflect the scale and complexity of the forest management operation. In the case of the Bainings, the assessment was undertaken without a prior scoping visit and used one SGS Forestry assessor. This reflected the scale of the operation and the fact that B&Q had already undertaken considerable field review and consultation to establish the probity of the management.

The assessment demonstrated that the Bainings Project was achieving its objectives of 'empowering' the local communities to control their own resources. For example, each village forms its own landowner company which holds the permit to cut timber. The income from the purchase of the timber goes directly to the village, which prioritizes the use of that income including bank payments for the sawmill, fuel and spares plus other essential expenditure. Training is provided to village members in sawmill use and maintenance and all decisions regarding sawmill siting, felling, species to be cut and allied matters are made by the Board of Directors of the Landowner Company or the village as a whole.

The assessment showed that these arrangements for managing the forest resource were currently appropriate to the Forest Stewardship Council's P&C. There was scope for further procedural tightening and ongoing research. However, such action was taking place. This was recognized as an important issue, particularly as the project expands into other village groups.

Source: Michael Groves, SGS Forestry, January 1995

Part 3

Current Initiatives and Views

---

9. Forest Initiatives and Certification

The crisis of worldwide forest loss and degradation has caught the imagination of environmental groups, government agencies, international aid agencies, electorates and others. As a result there are many international and national, governmental and non-governmental initiatives aimed at improving the sustainability of forest management. Many of these have helped to broaden the basis of concern for forests – from a narrower sustained timber yield goal at best, to a set of goals covering economic, social and environmental dimensions. A number of them are contributing directly or indirectly to the development of certification. Some key initiatives, and aspects of their work with implications for certification, are illustrated in Figure 9.1. They are further discussed below. Part 4 of this handbook includes a directory with details on international and national initiatives which are explicitly developing certification programmes.

International initiatives

Until the late 1980s, international initiatives led the way in generating guidelines and protocols for sustainable forest management. Some were broad, covering environment and development issues in general, but still having much to say about forests and the multi-sectoral roots of forest problems: notably the *World Conservation Strategy* (1980, revised 1991) and the *World Commission on Environment and Development* (1987).

There have been two initiatives of great significance for tropical forests: the International Tropical Timber Agreement [ITTA] which was set up in 1983 as a forum for consultation and cooperation amongst tropical timber producer and consumer countries (see Box 9.1 and Figure 9.2); and the Tropical Forest Action Plan (now Programme) [TFAP], which was set up in 1985 as a comprehensive action planning process and channel for development assistance activity (Box

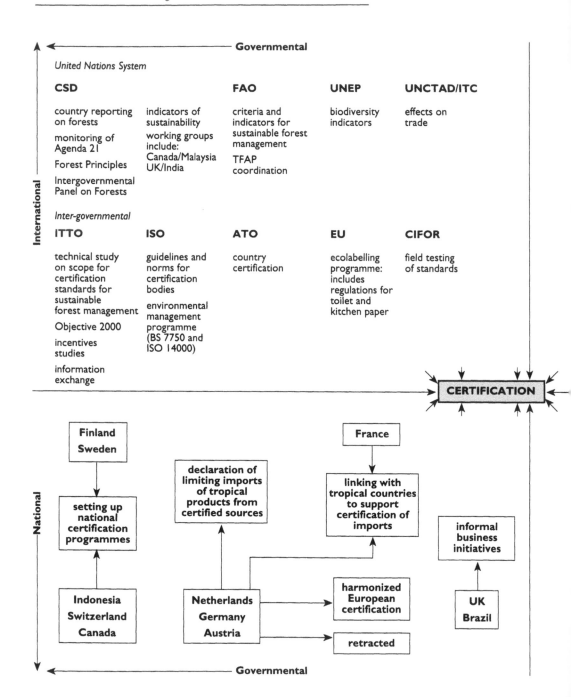

Figure 9.1 *Organizations contr*

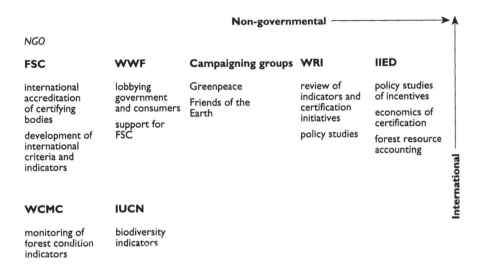

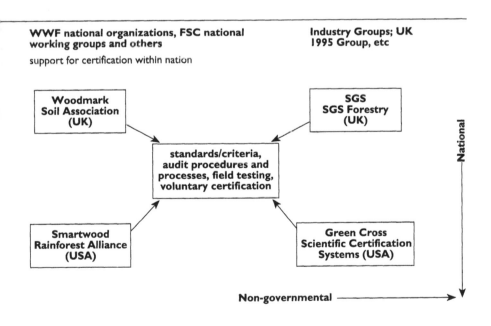

ing to development of certification

9.2). Since 1992, with the stronger realization that the forest problem was not at all restricted to the tropics, temperate and boreal forests have been emphasized in regional initiatives, described later.

At their best, these efforts have encouraged improvements in the way that national forest policies, institutions and procedures are set up. As a result, there is a better chance that varied local stakeholder needs for forests are taken into account. Commonly, however, they have failed for all the usual reasons that intergovernmental and development aid activities fail. In intergovernmental fora such as ITTA, forest issues became overly-politicized with the result that few countries were willing to move forward independently and make fundamental changes themselves.

In aid programme initiatives, such as TFAP, many project initiatives were not locally 'owned' and were too much for the country to absorb with local skills, staff and other resources. Their success became dependent on the presence of expatriate consultancy teams and equipment. Expatriates were often required to work

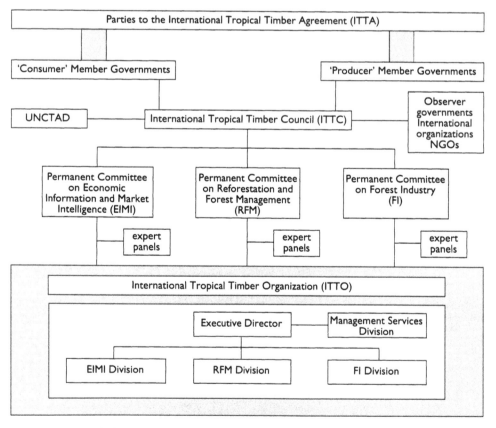

Figure 9.2 *Structure of the International Tropical Timber Organization (ITTO)*

within inefficient bureaucracies and were not held adequately accountable for their work. Such projects had a fixed short life and their efforts often fell by the wayside as they ended. Impatience and frustration with these initiatives led to more aggressive approaches. In the late 1980s, strong calls for bans and boycotts of tropical timber were made by some environmental groups. In North America and Europe, such calls were taken up by many local and municipal authorities and companies, and eventually some national governments such as Austria.

By the early 1990s, the possible major contributions of tropical forests to climatic stability, biodiversity and cultural heritage were being loudly trumpeted. This led to proposals for supranational control of tropical forests, often proposed by wealthier temperate countries but opposed by poorer tropical countries. By the 1992 Rio de Janeiro 'Earth Summit' – the UN Conference on Environment and Development [UNCED] (Box 9.3), many possible international initiatives, such as a global forests convention, were no longer possible. Even if the principles behind them were sound, poor tropical countries saw such initiatives as possible vehicles for yet more control by rich countries of their assets, and for acting as smoke screens to hide the environmental problems associated with the relatively high level of consumption of fossil fuels in rich countries.

One area where a global approach was deemed acceptable was in the transfer of additional finance from rich to poorer countries. The Global Environment Facility [GEF] was launched in 1991 to provide grants to developing countries for them to fulfil their obligations under the Biodiversity and Climate Conventions. The grants are for research, monitoring, and capacity-building and are administered by the World Bank, UNDP and UNEP. Major grants have been made for forest-related activities. Criticisms centre around the World Bank's top-down approach, weak project formulation with little participation, and increasing marginalization of environmental issues. In spite of the existence of GEF and other aid vehicles, net and total financial flows from rich to poorer countries are decreasing as a percentage of donor country GDP.

UNCED was highly significant in shaping the new institutional environment. As a result of the debates and agreements at UNCED, many contentious issues came to a head, and some were resolved. This has led to several significant trends in international objectives and approach:

- **All forests:** The issue is no longer just that of tropical forests. It is recognized that many temperate and boreal forests are being treated in unsustainable ways; the global extent of forest problems is generally agreed.
- **Social and economic issues:** Previously, environmental issues were considered as having top priority amongst forest problems at international levels. UNCED's emphasis on the need to combat poverty has raised the importance of both social and economic issues as part of sustainable forest management. Improved livelihoods for people living in and around forests, and the ability of developing country governments to increase revenues from existing forestry activities, are equally important aims.

Box 9.1 *International Tropical Timber Agreement/Organization [ITTA/ITTO]*

The International Tropical Timber Agreement (ITTA, 1983) aimed to establish a system of consultation and cooperation between tropical timber producing and consuming countries.

By the early 1970s, increasing incidence of boom-bust in the tropical timber market demonstrated the weakness of market information, planning and coordination systems in the tropical timber economy. This caused particular problems for – often poor – exporting countries. In 1976, the Fourth United Nations Conference on Trade and Development (UNCTAD) decided to develop agreements for individual commodities including tropical timber.

Over time, the perceived social and environmental degradation associated with tropical timber production was added to the agenda by campaigning NGOs. This meant that the eventual *ITTA (1983) was the first commodity agreement to deal with the potential conflict between development and environmental interests.* It covered the twin needs to conserve the resource base as well as to stabilize trade. These needs were reconciled by a Council ensuring decision-making parity between producer and consumer nations. The ITTA created the International Tropical Timber Organization [ITTO] as its organization and secretariat.

The ITTA was ahead of its time as a commodity agreement: it contained references to sustainable development, global resource issues, intersectoral linkages and intergenerational equity. In practice, aspirations for the ITTA have differed. The negotiators were learning about the problems as they went along, and the 'sustainable development' language was both comprehensive enough and weak enough to accommodate the possibility of inaction.

ITTO has generated political commitments to ensure that trade in tropical timber comes only from sustainably-managed sources by 2000 – 'Objective 2000'. ITTO has also agreed principles, guidelines and criteria on tropical forest and plantation management and on conserving biodiversity. However, members have had difficulty in translating all of these into action on the ground. Moreover, it has been argued that in focusing on industrial logging and the international timber trade, and on the relationship between national forestry agencies and the timber industry, many of the most important processes and actors in tropical deforestation and degradation were overlooked.

In the renegotiation of ITTA from 1994, the scope has remained very similar, despite arguments from Producer Members to extend coverage to all forests. They drew attention to the various links between tropical, temperate and boreal forests through issues related to environment and trade. The revised ITTA Preamble notes the desirability of implementing the UN Forest Principles [see Box 9.3, page 124]; and 'Objective 2000' has become a central focus, although not binding on Members; the text also states that timber prices should 'reflect the costs of sustainable forest management'.

ITTO remains among the foremost international fora for forestry and the timber trade. This is of great value – an intergovernmental forum is a critical need for continued progress on the international forestry agenda. ITTO was the first vehicle to take the intergovernmental aspects of forest certification seriously; in May 1994, ITTO completed an excellent study of the status of and views on certification. The subject was presented and debated at the ITTO formal meeting in Colombia in May 1994. No active role in certification was agreed for ITTO, however, except for information exchange and research. Since other international initiatives have been established, ITTO can focus more on resolving trade issues that affect sustainable forest management.

Box 9.2 *Tropical Forestry Action Programme (previously 'Plan') – TFAP*

The Tropical Forestry Action Plan (TFAP) was launched in 1985 in response to the escalating rate of tropical deforestation. It was a joint initiative by FAO, UNDP, the World Bank, the World Resources Institute and two bilateral donors. FAO assumed the international coordinating role. Under the motto 'saving the tropical forests', TFAP's initial goals were to curb tropical deforestation; to promote the sustainable use and conservation of forest resources to meet local and national needs; and to increase the flow of international aid to the forestry sector.

National Forestry Action Plans (NFAPs) are the country-level articulation of the TFAP. They are prepared and implemented by national Governments, normally with external donors and/or international NGOs. Most NFAPs are managed by a National Coordinating Unit, usually located within Government. Participation of national NGOs and the private sector is increasing, largely through the influence of NGOs such as IUCN, WRI and IIED. NFAPs are built around a sector review process, which ideally produces a long-term strategy, a policy framework, and a short-term action plan with priority project profiles. By the end of 1992, some 90 developing countries were active partners in the TFAP, and global donor assistance to forestry had quadrupled since 1985.

TFAP became a target of criticism – mainly as deforestation rates continued unabated – in spite of its efforts. By the end of the 1980s, deforestation was proceeding at an average yearly rate 50 per cent higher than in 1980. The expectations of many participants in TFAP, with regard to policy and institutional reform, to halting the agricultural and other non-forest causes of forest problems, and to increased financial flows, were not met. Some commentators accused TFAP of actually increasing some causes of deforestation. Or at least tackling them from an inappropriate angle defined by developed country interests – favouring industry over local people in an unaltered climate of policy and economic failure, where excess profits can be made from liquidating forests.

Many of TFAP's problems appear to stem from not taking adequate account of the root causes of forest problems; developing countries emphasizing sovereign rights to development; and developed countries emphasizing global environmental management. Several evaluations were launched in 1990. They indicated a need, since partially met, for more emphasis on environmental protection, the effective participation of local people, multi-sectoral analysis and development of cross-sectoral linkages, and less emphasis on long project 'shopping lists'. They called for a shift from being 'donor-driven' and 'project-oriented' to a longer-term, 'country-driven' and 'process-orientated' programme.

This shift has been taking place. There is more emphasis on strategic policy formulation involving interests beyond the forestry sector, and on local capacity-building. Operational guidelines have been prepared, as have useful guidelines on broader participation. NFAPs continue to be widely adopted as the principal planning and implementation framework for the forestry sector in many developing countries.

The TFAP is now a programme coordinated by FAO in Rome. However, reform of the TFAP's governance, management and contents has yet to be defined or introduced to everyone's satisfaction. Although there is a continued high demand for assistance through TFAP, political and financial support for the TFAP is waning. This is partly because of continued dissatisfaction at the low priority which FAO had tended to give forestry, an issue which FAO is now addressing with vigour.

TFAP has provided valuable experience of the operational and political aspects of mobilizing an international response to global deforestation. Many NFAPs are also undertaking some effective practical work, irrespective of the stormy climate of global forest politics.

123

> **Box 9.3** *1992 Rio de Janeiro 'Earth Summit' – UN Conference on Environment and Development [UNCED]*
>
> The United Nations Conference on Environment and Development (UNCED), held at Rio de Janeiro in June 1992, was the largest and most complex UN meeting ever organized. 120 heads of state were at the Summit which concluded the conference. Two major outputs dealt with forestry – the *Forest Principles* and the *Agenda 21* action plan for sustainable development. Both of them emphasize national-level decisions and actions. Three conventions were also the result of UNCED; and have a bearing on forests: the Convention on Biological Diversity, the Framework Convention on Climate Change, and the Convention to Combat Desertification. The Commission on Sustainable Development [CSD] was set up by UNCED as a UN body to monitor the progress of UNCED outcomes. It looked specifically at progress in the forestry sector in 1995 and recommended an Intergovernmental Panel on Forests to address key issues by 1997.
>
> Forests were amongst the most contentious issues covered at UNCED. Arguments were polarized, principally along a rich country–poor country divide. It was hard to distinguish between forests being discussed in a political context or politics being discussed in a forest context. Many *rich developed countries* argued that forests are of global importance, principally for biodiversity and climate regulation. They proposed that a degree of supranational control of forests is desirable, and many suggested a global, legally-binding forests convention.
>
> *Poor developing countries* stressed the sovereign right of countries to use forests for national development. They argued that global notions of sustainability could not encompass these varied and legitimate needs but, in contrast, more effective local control of forests was needed. Poor countries recognized that forests do provide global benefits. But they emphasized that, if there *were* to be a convention – an idea most were strongly against – a compensation mechanism would have to cover revenue foregone by countries setting aside forest reserves to serve 'global' interests. When it was clear that a compensation mechanism would not be forthcoming, nearly all vestiges of support for a convention disappeared in poor countries. Rather, the argument became that rich countries were seeking a forest convention as the cheapest way of obviating the need to cut down their own carbon emissions from high consumption and 'over development'.
>
> By the time of the Summit, most countries realized that there was little scope, and no time, to negotiate a legally-binding forest instrument. Rather than climaxing with a final clash over a possible convention, a positive approach – striving for a degree of consensus on forests – was sought by most.
>
> The result is the **Forest Principles** – the 'Non-legally binding authoritative statement of principles for a global consensus on the management, conservation and sustainable development of all types of forests'. They represent, *for the first time*, a global consensus on all forests, and not just those of the tropics. They constitute a political document, affirming general values.
>
> However, although aimed at international and national bodies, they are without clear operational links, and must be read in conjunction with Agenda 21, especially Chapter 11 on deforestation, but also several other chapters. It is the two together which form the 'first global consensus on forests' – and not (as the Principles state in Preamble d,) the Principles alone.

What the Forest Principles say:

The Principles cover all forest types and a wide range of associated environment and development issues. The 'guiding objective of the Principles is to contribute to the management, conservation and sustainable development of forests, and to provide for their multiple and complementary functions and uses' (Preamble b). The importance of local (forest) peoples and women is recognized, as is the need to support their 'economic stake' in forest use. The Principles note the need for valuing forests, setting associated standards, monitoring them, conducting environmental impact assessments for forest developments, setting aside protected forests, developing plantations, and developing national and international institutions with public participation.

The Forest Principles clearly put forward the view of poor countries – very much pro-development, stressing the importance of national sovereign rights over forests, and the need to develop forests sustainably in order to tackle both endemic rural poverty *and* environmental degradation. It implies the inappropriateness of supranational controlling roles; it contains statements emphasizing the importance of sovereign use of forests, but also refers to the need for additional financing.

Nevertheless, the Forest Principles are constructed of clauses which were either already part of national policies, or were ineffectually worded so as to provide escape clauses, or were without clear links to operational procedures. For 'progressive' nations they are weak; for less 'progressive' nations, there are enough qualifiers to justify any inaction on the Principles. (Other international principles and guidelines, eg on human rights, or even on tropical forest management through ITTO, tend to be constructed as 'high ideals' to be striven for; in contrast, the Forest Principles comprise only a set of lower common denominators which, collectively, challenge governments only a little.)

The outcomes of UNCED in general pointed to the national level as priority for organizing major change and responses. But 'national' is not necessarily the same as 'governmental'. UNCED stressed the role of civil society as a whole. Governments tend to do only what is in their own interests.

Technically, the Forest Principles could have made a clearer case for the integration of forest issues into land use planning, climate change, biodiversity, and socio-economic development. This is still necessary because other important UNCED outcomes – the Biodiversity and Climate Change Conventions – did not give advice specific to forests.

What next?

In summary, in the *content* of the Forest Principles there is little that is new; individual principles either reflect existing national policies or intentions, or are not considered relevant as obligations for specific countries. In its *form and political status*, the Forest Principles are clearly a precursor (or possibly a holding exercise) for further global consensuses in perhaps more specific areas.

Everything now depends on how the principles and action plans developed at Rio will be put into action – on how global bargains struck between rich and poor nations will be implemented. Some large, but slow-moving wheels, have been set in motion. Will the 'spirit' of Agenda 21 and the Forest Principles be followed as much as the 'letter'?

- **Quality of development aid:** Foreign aid is inevitably of limited quantity, and cannot be the sole means by which richer countries are able to assist poorer countries. The quality of development aid for forest problems is the issue. Aid should generally cover the incremental costs of the transition to sustainable development. It cannot bear the entire cost.
- **Recognition for NGOs:** NGOs have a recognized and often catalytic role to play in terms of brokerage between the different forest stakeholders.
- **Market-based incentives:** Incentives for sustainable forest management should be introduced as complements to regulatory approaches. The great size of the task to improve forest condition requires that the power of the market needs to be harnessed. There is recognition that the resources of the private sector are required.
- **Sovereignty:** It is accepted that major policy changes have to take place at the national level, respecting national sovereignty.
- **Criteria and indicators:** A great deal of work is going on, at political and technical levels, to define criteria and indicators and to clarify what has universal applicability and where local interpretation is required.
- **Harmonization and information sharing:** International initiatives are beginning to realize there is a need for concentration on harmonization, information-sharing and cooperation.

UN Commission on Sustainable Development

Existing institutions are evolving to match these new principles. UNCED intended that the UN system should undergo a major overhaul, but the only tangible change was in setting up the UN Commission on Sustainable Development [CSD]. A principal task of the CSD is to conduct obligatory assessments of the UN system's work, and to receive voluntary reports from nations. In different years specific issues will be addressed: in 1995 it was the turn of forests. One of the main outcomes of the CSD was the decision to set up an 'Ad Hoc Intergovernmental Panel on Forests', with the objective to promote multi-disciplinary action at the international level consistent with the Forest Principles. The main issues that the Panel will deal with are:

- identifying the underlying causes of forest problems and difficulties in implementing sustainable forest management;
- international cooperation in financial assistance and technology transfer;
- forest assessment, research and the development of criteria and indicators for sustainable forest management;
- an examination of trade issues, and their relationship to environmental issues;
- methodologies for full valuation of forests, with a view to promoting full cost internalization; and
- reviewing international institutions and instruments to further implement the Forest Principles.

The Panel will be composed of government representatives, with inter-governmental and NGO groups as observers. Its final conclusions will be presented at the CSD in 1997. It appears as if the Panel and the CSD will continue to provide the major UN fora for further developing consensus and action programmes on forest issues.

There are many other multilateral environment and development initiatives which have burgeoned since UNCED. For forestry, these are detailed below.

The Montreal Process

This now functions through the Working Group on Criteria and Indicators for the Conservation and Sustainable Management of Temperate and Boreal Forests. Since its first meeting in Montreal in 1993, a number of technical meetings have taken place between ten developed and developing countries. Seven criteria have been established (covering biodiversity conservation, ecosystem productivity, ecosystem health and vitality, soil and water conservation, global carbon cycles, multiple socio-economic benefits, and legal/policy/institutional frameworks); and work on indicators for each criterion began in 1994. This could provide a much-needed decision-making base for policy-makers, and pave the way for forest-level standards.

The Helsinki Declaration of the Ministerial Conference on the Protection of Forests in Europe (1993)

This Declaration aims to:

> *better contribute to national and regional objectives with respect to the rural sector, to the environment and to economic growth, trade and sustainable development in all European countries.*

The Helsinki process resembles the Montreal Process but is confined to Europe. It has a number of intentions:

*   to provide mutual assistance and coordination especially for forestry sectors in poorer European countries;
*   to reach agreements on data collection and reporting formats;
*   to research for mitigation of and adaptation to climate change;
*   to enable participation by local communities and NGOs; and
*   to stimulate generally the implementation of the Forest Principles, the Climate Change and Biodiversity Conventions and Agenda 21 in Europe.

It acknowledges European rural poverty, environmental degradation of forests and the limited resources of the forestry sector in Europe. Resolution H1 comprises General Guidelines for the Sustainable Management of Forests in Europe, which recognizes and builds on the Forest Principles and other UNCED instruments. In fact it does a reasonable job of combining the forest-related elements of these different instruments; and for the first time in a European document it defines sustainable management of forests. Resolution H2 comprises General Guidelines for the Conservation of the Biodiversity of European Forests; it recognizes and builds on the Biodiversity Convention and many other preceding agreements. These initiatives call for national planning and action without delay, and this has begun at the planning level.

But Helsinki also acknowledges the *global dimension* of European forests, which amount to approximately a quarter of the world's forests. One significant intention is to 'participate in, and promote, international activities towards a global convention on the management, conservation and sustainable development of all types of forests'.

The Intergovernmental Working Group on Forests (IWGF)

This was set up to bridge the gap between developed and developing countries on forest issues, particularly in view of the impasses at UNCED. A joint initiative of Canada and Malaysia, IWGF first met in April 1994, with representation from fifteen countries and a handful of international organizations and NGOs. More political than technical, its goal is to 'facilitate dialogue and consolidation of approaches', so as to identify elements of a possible consensus on *certain* forest issues mutually considered to be priorities. It has helped to resolve North–South differences. The final report was presented to CSD in 1995. A series of actions in seven categories has been agreed, with considerable consensus. The IWGF's working premises include national sovereignty; the use of forests for economic development and for their environmental roles; the need to apply the Forest Principles; and the need to complement other global forest initiatives and the climate change, biodiversity and desertification conventions. This potentially stands it on sound political ground.

The World Commission on Forests and Sustainable Development (WCFSD)

This is an initiative of prominent citizens, including Sweden's former Prime Minister, and inspired by the Brundtland Commission, headquartered in Geneva. It launched its rationale and possible mandate a year after UNCED, which covered putting forward a vision for forests and their sustainable development for the 21st century; periodically assessing the issues surrounding forests, and particularly the delicate politics of international forest issues; proposing means of strengthening sources of data on forests; proposing international cooperation and conflict resolution; and raising the commitment of all actors. It proposes Agenda 21 and the Forest Principles as its basis for action; and to offer a substantial input into the planned 1997 review of the Principles. Although it sought the patronage of the UN system, this was not supported by the UN Secretary General and WCFSD will be set up under the Interaction Council, a union of former heads of state. As such it has a limited political mandate, and will probably contribute more to debate over technical issues. The idea, of a select committee of 20–25 experienced and knowledgeable individuals assessing a highly dynamic subject, is a good one. However, the WCFSD could only ever be a complement to intergovernmental initiatives. For this reason, it is determined not to duplicate the new Intergovernmental Panel objectives.

Box 9.4 *What might a global forests convention achieve?*

There are three main issues concerning a global forests convention:

1   What are legitimate international roles in forestry?
These have been outlined in Chapter 4. To summarize, they are:

- Supporting national efforts at sustainability
- Supporting free trade in forest products
- Information sharing/cooperation
- Setting principles of sustainable forestry
- Standards harmonization
- Payments for 'global services'
- Encouraging and/or requiring national obligations to stakeholders
- International agreements on the above.

2   Which of the above cannot be achieved without a convention?
We suggest that it is only really *Payments for 'global services'*; and *Encouraging and/or requiring national obligations to stakeholders*. They are issues with considerable sovereignty implications and, for payment mechanisms, administrative implications – a good reason for dealing with them in the framework of a convention. But these are relatively new issues, about which there is little consensus; hence a convention is not yet politically feasible.

3   If a convention were to go ahead, how should it be set up?
- Scope should be global (not just tropical, for example); and comprehensive (balancing development, conservation, and different stakeholder interests)
- Encourage/oblige governments to address root causes of forest problems; and to reflect the needs of, and to provide incentives for, the different stakeholders
- Allow governments to define their own way to meet obligations
- Provide for funds/credit transfer (payments for global services)
- Clarify how to distinguish between global benefits and national benefits, in order to set payments
- Provide for arbitration
- Be legal in nature (but it is difficult to put the many different perspectives on forests into legal language)
- Have secretarial resources and a timetable for action.

One way to lead to such a convention is through first setting up regional conventions, such as currently exists for Central America.

The India-British Forestry Initiative

This was formed after UNCED to develop a reporting format for countries to share forest information on their progress towards Agenda 21 and implementing the Forest Principles. This was adopted as the preferred reporting format for the CSD's review of forestry in 1995. This format covers not only a country's review of progress, but also an analysis of problems and opportunities; it is therefore more forward-looking than the usual report to a UN body.

With this broad span of regional and international initiatives on forests, the notion of a global forests convention is again being raised, this time with some of the original antagonists proposing such an agreement. Box 9.4 introduces what a convention might eventually contribute.

The Forest Stewardship Council

This is a major initiative focused specifically on certification. It was born out of non-governmental concern about what was perceived as a failure of government husbandry over forest lands; it clearly sees certification as a significant response to forest problems. Composed of representatives from the scientific community, business, indigenous peoples and NGOs, it has been set up as an international non-governmental umbrella organization to accredit nationally-based forest certifying bodies. It will not be directly involved in certification activities itself, but aims to act as the ultimate accreditation authority. Box 9.5 and Chapter 11 provide considerable detail on this initiative.

The international NGO scene

This is becoming more sophisticated. Box 9.6 provides details on the 'big three': Greenpeace, WWF and Friends of the Earth. NGO styles of operating vary greatly, with traditional campaigning functions being supplemented to varying degrees by service and research functions. In essence, the approach of any NGO can be positioned somewhere on a line from a pure campaigning approach to a pure service approach. For most NGOs, the task of balancing campaigning, research and service to achieve their goals has become a critical issue.

For many NGOs, their main business is still to 'disagree': these are the campaigners, many of which, such as Friends of the Earth, have strong research and lobbying capabilities with many years of experience. They have been largely responsible for creating public awareness and are significant opinion-formers.

Other NGOs seek to reach compromise and to work with governments and the private sector; these speak the language of sustainable development – of participation and making trade-offs. These NGOs, like the International Institute for Environment and Development (IIED) and the World Resources Institute (WRI), tend to operate as specialist research centres. There is rather less experience, however, of this mode of operating, and it has meant that some groups have lost their

---

Box 9.5 *The Forest Stewardship Council [FSC]*

What is the FSC?

The FSC is an independent non-profit, non-government organization. It has been founded by a diverse group of representatives from environmental institutions, the timber trade, the forestry profession, indigenous people's organizations, community forestry groups and forest product certification organizations from 25 countries.

It has been established to provide consumers with reliable information about forest products and their sources. Public concern about forest degradation has resulted in a growing demand for those products which come from well-managed forests. This, in turn, has led to a proliferation of trade labels on timber products, and to the launching of independent certification organizations with different methods. FSC will help to eliminate confusing and false claims, and to give credibility to certification by accrediting those certifiers whose labels indicate that products have been produced according to FSC Principles and other nationally and internationally accepted standards.

The need for an accreditation agency, and the name of the Forest Stewardship Council, were first proposed in early 1990. A diverse group of concerned people from environmental organizations, the timber industry, certifiers and community forestry met in Washington DC in March 1992, and an Interim Board was elected. After several revisions of basic documents and ten national consultations ranging from Sweden to Papua New Guinea, the FSC's Founding Assembly was held in Toronto, Canada in October 1993. The Assembly was attended by 130 participants from 25 countries. It elected a Board of Directors, with a mandate to establish the FSC and to develop FSC's statutes and membership criteria and procedures.

FSC's Board is structured to achieve a balance between social, environmental and economic interests, as well as between North and South. Two places on the Board are reserved for representatives of economic interests, with strict selection criteria to ensure their independence from vested interests. The other seven Board members may represent social, indigenous, ecological, environmental, technical or other interests. Board members will have a tenure of three years.

The Founding Assembly decided to establish the FSC as an association, in which the members will be organizations (or, in a few cases, individual consultants). Government organizations may not be members of FSC, although their role in forest management as legislators and sometimes forest owners and managers is recognized. Apart from government agencies, the FSC is open to all organizations which support its objectives, and there is no limitation to their numbers of affiliations. The General Assembly of members will be the supreme authority of FSC. For voting purposes, the assembly will be divided into two voting chambers, one representing economic interests with 25 per cent of the vote, and one representing social, environmental and other interests with 75 per cent of the vote.

The Founding Assembly adopted the draft Principles and Criteria of Forest Management as a working document, pending some revisions which have since been completed.

The Board has selected Oaxaca City, Mexico, as the international headquarters of FSC and has appointed an Executive Director. The revised Statutes and Principles and Criteria have been approved by the founder members and their organizations in a postal ballot, by an overwhelming majority of the votes. The Board has the mandate to proceed with the legal establishment of the FSC, and to start accreditation of Certifiers.

What are the objectives of the FSC?
The FSC seeks to promote good forest management throughout the world, and will evaluate, accredit and monitor certification organizations which inspect forest operations and certify that forest products have come from well-managed forests. The FSC does not itself certify forest management or products; its mandate is to set a code of practice for certification, to accredit the certifiers, and to promote the development of national standards of forest management for the purposes of certification.

For accreditation, certifiers must demonstrate that they operate reliably, and that products carrying their labels have been produced from forests managed according to FSC principles and the recognized national standards where they exist. The FSC will continue to monitor the accredited organization's activities to ensure that standards are maintained.

The FSC principles, used for certification, are designed to ensure that forests of all types are managed in ways that are:

- environmentally appropriate;
- socially beneficial; and
- economically viable.

Environmentally appropriate forest management ensures that management practices maintain forests' biodiversity, productivity and ecological processes.

Socially beneficial forest management helps both local communities and society at large to enjoy long-term benefits from harvesting forests. Social benefits such as employment, revenue, useful products and guaranteed land rights provide strong incentives for communities to sustain the forest resources and adhere to long-term management plans.

Economic viability means that forest operations are managed so as to be sufficiently profitable, which enables stability of operations and commitment to good forest management. Economic viability does not mean financial profit at the expense of the forest resource, the ecosystem or affected communities. While recognizing the limitations of the marketplace, the tensions between the necessity of adequate financial returns and the principles of environmentally and socially responsible forest operations can be reduced through efforts to market forest products for their highest and best value, and through incentives (such as certification) which encourage investment in the forest resource.

These principles and associated criteria are intended to apply to all tropical, temperate and boreal forests. Many local factors will be taken into account when using them for evaluations of individual forests. Detailed provisions and interpretation will be covered by national standards.

FSC and accredited certifiers will not insist on perfection in satisfying the principles and criteria, but major failures in individual principles will normally disqualify a candidate from certification.

What is the FSC programme?

The FSC is starting its programme of evaluating and accrediting existing certifiers. Certification of forest products and management, and accreditation of certifiers, are entirely voluntary, but it is expected that most organizations wishing to gain credibility as independent certifiers will be interested in FSC accreditation. FSC procedures will follow the appropriate ISO Standards and guidelines for certification and accreditation. ISO is developing standards for environmental management, and FSC expects to collaborate with ISO and incorporate the standards in due course.

The FSC is also collaborating with diverse groups for the preparation of National Standards of Forest Management. They will be developed by consensus among interested and affected parties, including commercial, environmental, social, government and other interests, and should be compatible with FSC, ITTO, ISO and other recognized standards. As each country's standards are prepared and approved, they will be recognized by FSC and made an obligatory element of certification in that country.

FSC has been invited to collaborate in planning for certification and for national and regional standards of forest management in several countries, including Brazil, Costa Rica, Guyana, Indonesia, Mexico and the member countries of the African Timber Organization.

FSC's funding has come from governments and trust funds. In 1992–93, preparatory meetings and international consultations were funded by the governments of the UK (ODA), Germany (GTZ) and Australia, as well as WWF, the EC and the MacArthur Foundation. The costs of setting up the FSC HQ and of national consultations are funded by the governments of Austria and Mexico, with WWF-Netherlands and the Ford Foundation. These funds, paid even before FSC was fully constituted as a legal entity, demonstrate the widely-understood importance of certification, and the acceptance and credibility of the FSC.

In October 1992, FSC was invited by the Mexican Secretariat for Economic and Social Development (SEDESOL) to establish its headquarters in Mexico, which the Board of Directors accepted. The FSC office is operational in the city of Oaxaca. SEDESOL has offered financial support, and FSC has also been welcomed by the Sub-Secretary for Forestry and the Governor of Oaxaca. Staff of the federal and state forest services have agreed to collaborate with a national working group for preparing national standards. (See Chapter 11 for more details on FSC. Boxes 10.5 and 10.6 present arguments for and against the FSC.)

Contact: The Forest Stewardship Council
           Avenida Hidalgo 502, Oaxaca, Mexico
           Tel/Fax: +52 951 62110
Source: Timothy J Synnott, Executive Director, FSC

way in getting into the comprehensive agenda that sustainable development represents – finding that they are losing focus or competitive advantage.

Summary

Almost all of the international initiatives described above are concerned with: defining policies and principles for sustainable forest management; trying to get broad-based support from beyond the forest sector; and obtaining a balanced representation of rich and poor country interests, and conservation and development interests. However, each tends to place a different emphasis on the issues which surround sustainable forest management, as illustrated in Table 9.1, which covers some of those described in Chapter 5 and this chapter.

It is clear from this context that certification has to fit in with many initiatives if it is to be effective. A number of dimensions are being addressed by the new initiatives described above – but often it will be some time before results become apparent. In the meantime certification can firstly, contribute to the debate on solutions at national and international levels; and secondly, demonstrate solutions that work at a forest level in those places where certification is already taking place on the ground.

National initiatives

There have also been changes in national-level forest initiatives. In general forest authorities have, to a greater or lesser extent, widened their remit to cover such issues as recreation, landscape and biodiversity conservation. Many work more

Table 9.1 *International initiatives showing the main sustainable forest*

International initiative	national legislation	forest health and vitality	forest productivity	biodiversity conservation	soil & water conservation	tenure rights	local people rights
FSC	***	*	***	***	***	***	***
CSD		*	*	**	**		*
UNEP				***			
ITTO	***	***	***	**	***		
ISO/CSA	***	*	**	**	**	**	**
EU	***	***	**	**	*	*	**
IUCN				***	**		
CSCE	***	***	**	**	*	*	**
HELSINKI	***	***	**	**	*	*	**

Key:
FSC – Forest Stewardship Council
Ten Principles and Criteria of Forest Management
CSD – Commission on Sustainable Development
Indicators of sustainability - forestry format currently being worked out
UNEP – United Nations Environment Programme
Biodiversity indicators
ITTO – International Tropical Timber Organisation
Criteria for the measurement of natural forest and plantation management in the tropics
ISO/CSA – International Organisation for Standardisation/Canadian Standards Association
Standards for environmental management systems [ISO 14000] for forestry [CSA]

in partnership with different stakeholder groups, reflecting the increasing need to accommodate the wider aspects of forest use.

Partly in response to NGO and other international initiatives there is also a number of national initiatives which explicitly recognise certification. These national initiatives fall into three broad types:

1 development of national certification programmes;
2 use of certification to identify traded wood – mainly tropical wood – as originating from sustainably managed forests; and
3 initiatives which aim to encourage certification with main trading partners and suppliers.

Setting up national certification programmes inevitably involves achieving a consensus between at least the majority of stakeholders. The most active of these initiatives in Switzerland, Sweden, Indonesia, Finland and Canada are all striving to achieve this [see Chapter 12]. In several of these countries, detailed draft documentation has now been produced which is expected to be finalized during 1995. The CSA initiative in Canada has so far produced the most comprehensive set of materials. See Box 9.7.

management criteria covered and their relative emphasis within each initiative

work safety	*appropriate documentation*	*appropriate skills and training*	*monitoring and assessment*	*environmental management systems*	*prioritisation of environmental effects*	*targets and continual improvement*
***	**	***	***	*		*
*	**		**			
***	***	***	***	***	***	***
**			**			
**			**			
**			**			

EU – European Union
Ecolabels for toilet and kitchen tissue paper including criteria for forestry
IUCN – International Union for the Conservation of Nature
Biodiversity indicators
CSCE – Conference on Security and Cooperation in Europe
Boreal and temperate forest indicators
The HELSINKI Process – Ministerial Conference on the Protection of Forests in Europe
Boreal and temperate forest indicators

(***=strong emphasis; **=medium; *=weak)

Box 9.6 *Campaigning environmental organizations: 'the big three'*

Greenpeace	Amsterdam, Netherlands [international HQ]	tel: +31 20 523 6555 fax: +31 20 523 6500
	London, United Kingdom	tel: +44 171 354 5100 fax: +44 171 696 0012
	Washington DC, USA	tel: +1 202 462 1177 fax: +1 202 462 4507

Origins: Established in Vancouver, Canada, in 1971 as a focus for campaigns against nuclear testing in the Pacific. Greenpeace International was set up as a coordinating force in Amsterdam, 1979. Greenpeace has 4.5 million members in 158 countries [1.6 million in the USA].

Aims: Greenpeace is: 'dedicated to the protection of the environment through peaceful means ... Greenpeace embraces the principle of non-violence and rejects the use of physical force against people or property'.

Activities: Greenpeace's main campaigning areas are: climate change, the ozone layer, nuclear issues, chlorine, toxic waste, oceans, forests and Antarctica. Greenpeace Russia, set up in 1992, is the main campaigning group focusing on the threat to Siberia's forests from uncontrolled logging.

WWF [World Wide Fund for Nature]	Gland, Switzerland [international HQ]	tel: +41 22 364 9111 fax: +41 22 364 3239
	Godalming, United Kingdom	tel: +44 1483 426 444 fax: +44 1483 426 409
	Washington DC, USA	tel: +1 202 293 4800 fax: +1 202 293 9211

Origins: Established in 1961 in response to the decimation of wildlife; WWF has over 5 million members worldwide. The organization consists of 23 National Organizations and 24 Programme Offices worldwide.

Aims: WWF's core is conservation. There are three basic aims: (1) preserving biodiversity; (2) reducing pollution and the wasteful consumption of resources and energy; (3) ensuring the sustainable use of natural resources.

Activities: WWF is also a service organization: it is for 'dialogue not defiance' – such as involving corporate sponsors in a commitment to continuous improvement of environmental practice. As of 1994, over 10 000 projects in 130 countries, ranging from well-focused political campaigns on wildlife trade to efforts – often working directly with governments – aimed at protecting particular habitats. WWF embraces the concept of sustainable use, with an emphasis on the needs of local people.

FoE	Netherlands	tel: +31 20 622 1369
[Friends of	[international HQ]	fax: +31 20 639 2181
the Earth]	London,	tel: +44 171 490 1555
	United Kingdom	fax: +44 171 566 1655
	Washington DC,	tel: +1 202 783 7400
	USA	fax: +1 202 783 0444

Origins: Set up in 1971 and, as of 1994, established in 52 countries with a combined membership of one million.

Aims: Are wide ranging and include: protecting the earth against further deterioration; preserving ecological, cultural and ethnic diversity; increasing democratic decision making in the protection of the environment first and foremost by the people most directly affected; and achieving social, economic and political justice, and access to resources and opportunities at local, national and international levels.

Activities: embrace: climate change, acid rain, the ozone layer, energy and nuclear power, water quality, transport, agriculture, rainforests and biodiversity; encompassed within the wider politics of consumerism, trade and sustainability. FoE's work on forests has consistently offered solutions – such as the 'Good Wood Guide' – and researched the issues in detail.

Some form of 'certificate of origin' to identify wood as coming from sustainable sources has been suggested to varying degrees. In Austria, legislation was enacted and subsequently retracted; both Switzerland and the Netherlands have considered such options. However, in the case of Switzerland the option was only to identify the country of origin. The difficulties which such programmes have faced, in terms of the practicality of tracing products, and in dealing with accusations of unfair and discriminatory trading practices, make them difficult to implement.

Initiatives which aim to encourage the production of certified wood are either the result of private-sector initiatives, or have some form of government sponsorship.

The clearest examples of private-sector initiatives are in the UK [the 1995 Group] and Brazil. The Swiss national programme also aims to provide support for the development of certification in poorer countries.

Government-sponsored national initiatives include those in Germany [*Projekt Tropenwald*], the Netherlands and France. The initiatives in Germany and the Netherlands are more advanced than that in France, and also include significant NGO and private sector involvement. Many of these countries' trading partners are wary of certification, and react cautiously to the approaches made. One reason is poor understanding of the implications of certification for the performance of the trading partner's forest industry and forestry sectors.

Further details on these active national initiatives are provided in Chapter 12.

Box 9.7 *The Canadian Standards Association – Sustainable Forest Management System*

The CSA standard for Sustainable Forest Management (SFM) has grown out of the need to provide assurances that Canadian forests are being managed to an acceptable standard. The standard follows an environmental management system format – hence the title SFM System.

The Canadian Council of Forest Ministers – CCFM (1992) – set the goal of SFM as '*to maintain and enhance the long-term health of our ecosystems, for the benefit of all living things both nationally and globally, while providing environmental, economic, social and cultural opportunities for the benefit of present and future generations*'. The standard also reflects agreement in Canada that, in the context of sustainable development, a major objective of forest management should be to maintain native biodiversity to the fullest extent possible.

There are three essential inputs into an SFM System which can be audited:

* the current state of the forest (as defined by existing conditions and uses);
* stakeholder input related to the forest; and
* the management goals and objectives for the forest.

These three inputs, typically unique for each forest, dictate how the components of a SFM system are applied. The components of a SFM System, designed to manage the above inputs, are:

1 commitment;
2 public participation;
3 planning;
4 implementation;
5 measurement and assessment; and
6 review and improvement.

Each component is subjected to audit procedures which assess their conformance with certain principles and criteria. Principles include:

1 an open and transparent process that allows for stakeholder participation;
2 a requirement for clear focus and planning;
3 identification of criteria and indicators based on those produced by the CCFM;
4 allocation of sufficient resources to achieve its SFM objectives;
5 a requirement to continuously measure, monitor and assess performance; and
6 continuous review of, and improvements to, the SFM system.

The criteria developed by the SFM, apart from satisfying the level defined by the CCFM, should contain the following elements:

* Conservation of biological biodiversity: ecosystem diversity, species diversity, genetic diversity;
* Maintenance and enhancement of forest ecosystem condition and productivity: incidence of biotic and abiotic disturbance and stress, productivity;
* Conservation of soils and water resources: physical environmental factors, policy and protection forest factors;
* Multiple benefits to society: long-term sustained yield, maintenance of habitat and animal population levels, availability and use of recreational opportunities; and
* Accepting society's responsibility to sustainable management: sustainability of forest communities, fair and effective decision making.

Note: The above information is taken from the February 1995 version of the CSA SFM System. The documentation is currently in draft form.

Box 9.8 *The dangers of self-certification*

In response to green consumerism, companies are eager to promote the green credentials of their products. This is apparent in many sectors, particularly in the wood and paper industry, as consumers are making the link between the timber trade and the destruction of forests around the world.

Labels abound on many items, carrying a range of claims such as: timber coming from a 'sustainable source' or from 'renewable forests'; 'for every tree felled three more are planted'; or even 'buying this product will help save the rainforests'. Though the wording may differ, the message is the same, with the implication that the wood or paper product has come from a well-managed forest. Customers are asked to rest assured that their purchase will not have caused forest destruction.

In 1990, the World Wide Fund for Nature UK (WWF UK) decided to investigate the reliability of such claims. The study concentrated on the tropical timber trade, and found that half of the companies studied were willing to make some assurance to customers regarding environmental acceptability. A more in-depth look at 81 of these companies found only three of these willing or able to make any attempt to substantiate their claims, and even those companies were unable to answer fully the simple questions posed by WWF UK.

The nature of claims varied considerably, from the factually inaccurate to those which, while not actually false, were certainly misleading. The phrase 'from a managed forest', for example, means little but may well convince a consumer that all is well with the product in question. Faced with a bewildering array of claims, it is extremely difficult for concerned consumers to make an informed choice.

Further research commissioned by WWF UK showed that many retailers, merchants or importers simply passed on information supplied to them by timber trade associations or timber exporters.

In many cases, *certificates of 'sustainability'* accompany shipments of timber, reflecting the response of producer countries to the threat of timber boycotts. The certificates appear to provide documentary evidence that the timber has come from a sustainably managed forest. Certificates from the Ghanaian Forestry Department state: '*We confirm that all Ghanaian tropical hardwoods supplied by [company name] come from forest resources which are being managed to ensure a sustained yield of timber and other forest products in perpetuity and to arrest forest depletion and environmental degradation*'. As well as government certificates, companies often 'certify' their own timber. A company operating in Sabah, Malaysia, provides a 'Certificate of Products from Sustained Yield Management'. This 'confirms' that 'the hardwood rainforest products supplied come from well-managed production forest in accordance with the principle of sustained yield management thus safeguarding the environment and the ecological balance'.

As well as such 'certificates', many countries have launched ambitious promotional campaigns to counter adverse publicity. Such publicity work is not restricted to tropical countries. Alarm over temperate forest mismanagement is increasing, and many temperate producers have responded by aggressively marketing their timber.

The rise in the use of labels and advertising materials, which attempt to dispel environmental concerns, raises a number of issues concerning self-certification. In particular, *country certification*, thought by many to be a simpler route to improved forest management than certification of the forest management unit, has a number of failings:

- Country certification would allow 'free riding' of poor forest managers within certified countries, and provides no incentive to the individual producer to improve practices within non-certified countries.
- Although national forest policies and guidelines may outline a tight framework for forest management, the reality of logging practice is often a long way from this ideal. Without inspection of management in the forest, the reliability of certificates of sustainability will be suspect. The standard of forest management will vary widely within a country. Country certification would offer little incentive for the individual operator to improve performance. In some countries, even if logging operations do adhere strictly to the law, they could not be considered sustainable. This point was emphasized by Simon Rietbergen in *No Timber Without Trees* [Poore et al (1989), Earthscan Publications, London], when referring to concession agreements in many African countries: *'quite a few of the elements that need to be part of the logging process, if forest management is to be sustainable, are not even mentioned in the present concession agreements'*. Clearly government guidelines cannot be used as a guarantee of sustainability.
- Without accurate tracing of timber, any environmental claims must be open to question. This is especially relevant to country certification, where there would be an incentive to route timber through 'approved' countries.

All self-imposed certification schemes lack one vital component – independence. A credible labelling system must not have commercial interest in the outcome of monitoring of forest management. Any system in which the 'certifier' stands to benefit by approving forest management cannot be considered objective.

In conclusion, the most promising route to use the power of the market to improve forest management is independent certification at the forest level. Such a system avoids the pitfalls outlined above, and would enable producers making the greatest progress towards sustainability to be rewarded directly.

Main source: 'Truth or Trickery' by Mike Read, WWF [UK], 1991

Initiatives which are commonly confused with certification

The demand from wood buyers to provide them with more information regarding the origin of wood and paper products, and to demonstrate the sound management of the forests from which they derive, has spawned many initiatives. Several, however, make claims which are not backed by a credible certification programme. These claims either tend to concern the management of forests from which wood and paper products come; or are national labels promoting the generally good forest management of a particular country. Neither of these types of initiative should be classified as a certification programme. In particular, many of the claims lack the vital ingredient of independent, third-party verification. Even where such initiatives include some degree of monitoring or control, the result is almost always self-certification.

Box 9.9 *National labelling programmes to certify origin*

National labelling programmes, which essentially consist of a label verifying the country of origin, are not to be confused with certification. They are based on the premise that, in the country concerned, forests are generally well managed; thus any purchaser should be confident that the labelled timber has been produced in accordance with good forestry practice. Inspection of timber product companies is conducted in order to verify the *origin* of wood – and no more. There is no inspection of the forest or its management, and the label does not make any claims about 'sustainability'. Such labelling schemes have been set up by national forest industry organizations in order to promote home-grown timber. They offer no incentive for change within a country.

Three examples of national labelling programmes from Europe are briefly described:

Certificate of origin 'Swiss Wood'
In 1988, nine timber trade and industry organizations agreed to the creation of a certificate of origin for Swiss wood. The *Comité du Bois Suisse* [CBS] was appointed as secretariat to administer and enforce the rules of the scheme. The scheme's objective is to increase harvesting of Swiss forests by promoting the demand for Swiss wood – the premise being that Swiss forests are in general well managed – facts which should be identified and communicated to consumers.

CBS gives the authorization to use the designated label on wood products and in promotional materials. The authorized companies have to use as much Swiss wood as possible. Individual products which carry the Swiss Wood label must be 100 per cent from Swiss forests for logs and primary processing; and 75 per cent from Swiss forests for materials from secondary processing. The visible surface of finished products should not be plastic, nor of a wood not grown in Switzerland.

Once initial authorization has been given, use of the Swiss Wood label relies on goodwill. In the case of complaint, CBS has the right to external inspection, using experts which it appoints. In the case of misuse, the right to use the label is removed and company concerned is deleted from the list of approved organizations.

The 'Plus Forest' project in Finland
The Plus Forest project was set up in 1992. Plus Forest is a Finnish market-oriented communications programme. It is also a brand name for Finnish forestry and forest industry products made of Finnish wood. The project was founded mostly as a reaction to the environmental pressures originating in export markets and Finnish environmental groups.

The Finnish forest industry is export-oriented, and therefore concerned with informing customers of the acceptability of Finnish forest production. On an international level, Plus Forest tells the purchasers of Finnish wood and paper products about general economic, social and ecological dimensions of wood production in Finland.

The Finnish Forestry Association controls and coordinates the project. The central idea in the Plus Forest project is to link all aspects of forestry together under a common symbol. The forest industry, woodland owners, forestry professionals and government authorities are all promoting the same cause.

Plus Forest is not an ecolabel. It is a brand name for Finnish forestry. It is based on a series of principles, agreed by the forest industry and forest owners.

The Forest Industry Council of Great Britain [FICGB] 'Woodmark'
In June 1994, the FICGB introduced 'Woodmark' [not to be confused with the Soil Association sustainable forestry label of the same name].

Woodmark is intended to help specifiers, buyers and users of British timber to inform themselves as to where the timber came from, how the forests are regulated, and what the regulations are. They are encouraged to visit the forest and see for themselves and so make their own informed choice.

Woodmark guarantees that at least 90 per cent of the wood product so labelled comes from British timber, felled in accordance with prevailing government regulations. This purely factual information can, upon challenge, be proven through production of the Bill of Sale, the Supply Contract and the original Felling Licence or Plan of Operations.

The scheme refers to current Government regulations, and by doing so invites scrutiny of these regulations. It responds to an undertaking given by British Ministers at Helsinki (1993), and subsequently repeated, to publish a summary of forestry regulations and controls for ready public reference, backed up by the detailed guidelines and codes used within the industry.

Woodmark also emphasizes the function of felling licenses and permission. This is because in Britain felling permission is compulsory (except in a very few exceptional circumstances); because felling licenses usually include conditions regarding restocking and future management; and because the Forestry Commission has strong legal powers of enforcement. Hence this is a practical and effective point of intervention.

The FICGB Woodmark does not make any claims that it is an informative label. The implication is that, because the timber is British, it must have been produced in accordance with the Forestry Commission regulations – which in turn must mean that the forest was well managed.

The 'Made in Country X' labelling initiatives do not necessarily make direct claims concerning particular wood or paper products; but are rather designed to certify product origin, and to imply that forest management in the country is generally of a high quality. Details of three such initiatives, in Switzerland, Finland and the UK, are given in Box 9.9. These initiatives suffer the same problems as proposals for country certification – such as that promoted by the African Timber Organization [ATO]. Whilst theoretically country certification should be possible – see Chapter 7 for more on this – these national label programmes also tend to exclude possibilities for independent third party verification.

WWF [UK] has investigated the reliability of various claims on wood and paper products and produced a report called 'Truth or Trickery'. The findings of this work are summarized in Box 9.8. The unwillingness of such initiatives to accommodate independent verification of their claims, together with the wide array of labels used, serves to reduce their credibility and effectiveness. Credibility is low, because buyers do not always trust the authenticity of the claims made; and effectiveness is low, because the consumer is unable to make informed purchases amidst the not insignificant number of differing labels.

10. Views on Certification

This chapter presents a range of current views on forest certification. The main points raised have been summarized and some typical contributions quoted in boxes. The material for this chapter has come from a variety of sources which includes:

- direct consultation by the authors with producers, trade associations, retailers, environmental NGOs and professional foresters in both temperate and tropical countries;
- governmental insight, obtained principally from the public record of ITTO members during the debate on certification at Cartagena in Colombia [May 1994]; and
- other viewpoints collected from recent published statements given by interested parties.

As has been pointed out in earlier chapters the idea – and especially the practice – of forest certification is young. It is entering an arena populated by many other types of initiative aimed at solving forest problems. There is still scope for certification to be targeted more appropriately, and for potential benefits to be more carefully researched and described. By setting out the broad spectrum of views, it becomes clear that there are many issues for which a consensus should be achieved if certification programmes are to be successful.

To date, limited information on certification has been available to help stakeholders form their views, especially in developing countries and among minority stakeholder groups. Few studies and consultations concerning the implications of certification have been made. Many views tend, therefore, to be general or based on limited experience. Regular articles appear in professional and trade publications and numerous conferences have been organized, either on certification in particular or with certification as part of the proceedings. Many government views have been informed through the report (Ghazali and Simula, 1994) presented to the ITTO working party on certification; and subsequent discussion in Cartagena. Other views have been formed with respect to the FSC initiative. The FSC national consultation process during 1993 was particularly helpful for getting many groups thinking about certification.

Certification, both from conceptual and operational viewpoints, has been and continues to be a learning process for many stakeholder groups. Many groups have been obliged to work out a position from first principles, which not all are equipped to do. In particular, nearly all stakeholder groups have difficulty with the structuring of certification programmes as set out by ISO. Definitions also present particular problems. There is frequent confusion, for example, between certification, eco-labelling, chain of custody inspection, life cycle analysis, and management information systems.

Nonetheless, there has been a huge change in the debate from a position in 1991 when 'certification' was new; the knowledge base was low; and governments in particular were reluctant to even consider certification as a potential tool. It is our belief that continued discussion and consultation will further this process and, should certification prove to be a useful tool, programmes will be produced which have significant practical application. An important product of this process will be the building of trust between the different stakeholder groups. This is likely to have other, wider benefits to the local community beyond improved forest management.

Views of the main stakeholder groups

We have grouped stakeholders into five main groups: governments; environmental NGOs and independent observers; indigenous peoples' NGOs and social groups; forest industry and the timber trade; and consumers and retailers. A selection of views from these groups are presented in Boxes 10.1 to 10.14 (pp155–67); and a list of these views by box number and stakeholder group is given in Table 10.1. We have selected views which, in our opinion, represent the main issues and positions of each stakeholder group. A relatively large number of views are presented for stakeholder groups where a broad spectrum of views is present. This is most noticeable for the group: 'environmental groups and independent observers'. By contrast, consumer and retailer group organizations have basically similar views on certification which are summarized well in the submission from Paul Ankrah at Do-it-All, a UK-based 'do it yourself' retailer.

Table 10.2 summarizes what appear to be the current views of stakeholders following our grouping. All groups seem to agree that certification should cover at least the following:

- environmental issues;
- include international consensus and clarity; and
- require certifying bodies to be accountable rather than self-appointed.

Table 10.1 *Summary of views presented on certification (see Boxes 10.1–10.14 for details)*

Governments	Environmental NGOs and independent observers	Stakeholder group: Indigenous peoples NGOs and social groups	Forest industry and the timber trade	Consumers and Retailers
Malaysia delegation to ITTC	**A view from the "South"** R de Camino, C Sabogal, PJ Martins	**World Rainforest Movement** Forest peoples group	**UK** Forest Industry Committee of GB	**Do It All – UK** DIY retailer
UK Overseas Development Administration	**Pro Regenwald – Germany** An environmental NGO	**Soltrust – Solomon Islands** Conservation NGO	**Alpi – Cameroun and Italy** Concession holder and manufacturer	
Canada Canadian Forest Service	**WWF – UK** An environmental NGO		**Aracruz – Brazil** Forest owner and pulp producer	
	Soil Association – UK An environmental NGO and certifier		**Netherlands** Timber Trade Association	

Table 10.2 *Summary of stakeholder group positions on certification issues*

Stakeholder group	Governments	Environmental NGOs and independent observers	Indigenous peoples' NGOs and social groups	Forest industry	The timber trade	Consumers and retailers
Principle of certification	+ + −	+ + +	+ + −	+ + −	− − +	+ + +
Independent audits	− − −	+ + +	+ + +	+ + −	− − +	+ + +
Country level certification	+ + −	− − −	− − +	− − +	+ + −	− − −
Forest level certification	− − +	+ + +	+ + −	+ + −	− − −	+ + +
Environmental issues included	+ + +	+ + +	+ + +	+ + +	+ + +	+ + +
Social issues included	− − +	+ + +	+ + +	− − +	− − +	+ + −
International consensus and clarity	+ + +	+ + −	+ + −	+ + +	+ + +	+ + +
Certifier accountability	+ + +	+ + +	+ + +	+ + +	+ + +	+ + +
Transparency	+ − −	+ + +	+ + +	+ − −	+ − −	+ + +
Allowance to improve over time	+ + −	− − +	− − +	+ + +	+ + +	+ + −
Process focused	+ + −	− − +	− − +	+ + +	+ + −	− − +
Event focused	+ + −	+ + +	+ + +	− − +	− − +	+ + −

Key:
strongly for   + + +
moderately for   + + −
moderately against − − +
strongly against   − − −
Note: Survey conducted during 1994

Quite how such accountability should be achieved is, however, the subject of very different opinions.

The greatest differences between groups concern the need for independent third party verification [audits]; country-level certification as opposed to forest-level certification; the extent to which social issues should be included; and the extent to which forest operators should be given time to adapt within a certification programme. This last point is also wrapped up in the potential trade discriminatory effects of certification; and whether certification should be focused more on forest management 'processes' or on 'events'. The discussion in Chapter 7 demonstrated the importance of good management systems in certification.

Individual stakeholder group views are summarized below.

Governments

Governments vary in their support for certification. Most now believe that certification, or something like it which will involve third-party verification of forestry activities, is inevitable.

Analysis of the discussions at the International Tropical Timber Council [ITTC] can be used to help register shifts in governmental opinion. These have moved from a position where many governments did not wish to consider certification at all, believing it to be a trade barrier (in 1992), to one where consultants (Ghazali and Simula) were appointed to marshall the facts and opinions (in 1993), to one where governments were sharing their ideas on principles and prerequisites for certification [in 1994].

There was extensive discussion of the potentials of certification at the ITTC meeting in Colombia, May 1994. Many governments rallied around the well-articulated Malaysian position [summarized in Box 10.1].

Today, most governments offer *tentative support* to certification, or at least they accept the notion of certification. Such governments tend to list a number of conditions for certification to be acceptable: notably non-discriminatory practices and principles; the need to phase certification in; the need – or otherwise – to allow conversion forests to be certified; the degree of international control/monitoring; the type of external body deemed acceptable; the need for national coverage; and the need or otherwise for external funding to cover the costs of a transition to certification.

Some governments also tend to raise some fundamental doubts about certification. Opinions vary, however, depending upon which government department is consulted. A principal doubt is whether it is possible to enforce certification, in the sense of keeping uncertified timber out of timber flows. Others are the practicality of assessing sustainability in the field; and the extent to which certification – as a trade instrument – can be as effective and direct a tool for sustainable forest management as other policy measures. The US federal government, for example, doubts whether this is the case, unless certification becomes very widespread – and then it doubts that certification organizations alone will be adequate to institutionalize certification.

Opinions also vary as to the scale and scope of certification. With regard to scale, most debate is over whether 'country certification' should be allowed for – which many governments would prefer. With regard to scope, most governments accept that certification should cover environmental issues; but there is particular debate over the extent to which social issues should be included, especially those that concern the rights of indigenous peoples, and whether social impacts wider than 'first-order' (neighbouring community) impacts should be included.

Governmental support for certification tends to come more firmly from importing governments [such as the Netherlands and Austria], and from governments of countries which export to 'greener' markets in Western Europe and North America [such as Indonesia, Finland and Sweden]. Some African governments, for example, see certification as helping to improve the 'ecological competitiveness' of African timber.

Environmental NGOs and independent observers

Environmental NGOs are generally supportive of certification, but views are quite varied, reflecting the diversity of interests that this group represents. European and North American NGOs that campaign on tropical forest issues have more strongly-formed opinions than others; whereas some groups in tropical countries have barely considered the issue. Nearly all members of this group are strongly against country-level certification.

By and large, environmental NGOs claim that traditional regulatory and aid programme approaches to solving forest problems have failed – one of the reasons why such groups tend to be cautious of government- or industry-inspired certification initiatives. Of the alternatives, most tend to prefer certification to other more 'short, sharp' approaches such as bans and boycotts. Many NGOs that are in favour of forest certification appear to presume that certification will directly solve major forest problems; but they do not present a rationale as to why or how. Some pro-certification NGOs have a self-interested perspective, seeing themselves as potential certifiers.

Many environmental NGOs have supported the FSC initiative, with some – such as WWF – playing a leading role in it. Even those groups that have reservations about FSC are, in the main, attempting to improve the initiative, such as how FSC deals with social and participation concerns, rather than campaigning against FSC.

Other environmental NGOs, however, believe that commoditization and corporate approaches to forests are intrinsically anti-forest and anti-forest peoples – because of the power structure of big business and prevailing short-term profit-seeking attitudes. They believe that certification will be coopted by the prevalent economic interests; although some note that certification may work if is part of a regulatory – rather than a purely voluntary – approach. They see the move to 'process-focused' certification solutions, such as an environmental management systems approach to certification, as an attempt by industry to bypass the need for absolute, high standards which should define sustainable forest management.

Indigenous peoples' NGOs and social groups

These groups, particularly from tropical countries, are generally less convinced than environmental NGOs that certification will be of assistance. Otherwise, their views are broadly similar. Local people with forest interests are often concerned that their land rights and aspirations for economic development will be ignored by the certification process. The focus that many certification initiatives have on environmental issues, together with the fact that they are often driven by urban interests in richer countries, makes many in this group fear that certification will result in a 'locking up' of their forest for reasons of global conservation and/or corporate forestry. Indigenous groups are concerned that certification programmes may pay insufficient attention to issues of land rights and local livelihood and development needs; and thereby may legitimize the practices of current forest users with whom local people are often in conflict.

Forest industry and the timber trade

Trade and industry views are the most varying. As there is sometimes a significant difference in views between forest industry and timber trade groups, we have shown these separately. Table 10.2 shows that these differences are mostly over the principle of certification; the need for independent audits; and whether or not certification should be at a country or forest level. In general, views tend to be more positive the closer to the 'green' consumer the organization is. Members of this group also share many views in common. Primary among their concerns is high or unnecessary cost, especially the cost of something that on the face of it does not appear either to offer any commercial advantage, or to be practicable. The views tend to be formed around issues such as loss of flexibility in supply, because of the need to guarantee chain of custody; the need for attention on processes rather than events; and the need to allow for time in changing forest operations to meet the demands of sustainable forest management standards – forests take a long time to put right. This group is more adamant than most on the need for a level international 'playing field', so that everyone knows where they are; and for certifying bodies to be accountable, so that the industry and trade will not in the future be held hostage to certification programmes dominated by environmental NGOs.

Those who are trading within, or importing into, markets with increasingly strong 'green' preferences, see future benefits and are in strong support. Often traders in these same markets are reacting defensively to what is perceived as an inevitability. There are others, often new entrants from exporting countries, who want to improve market access and share; and to create market niches for specialized products. Many of these, however, have been disappointed at how difficult this is. It appears that even certified timber is difficult to market unless the price is right, and delivery schedules and consistent quality can be guaranteed.

In some countries, forest industry and the trade are protagonists for certification. In Australia, the National Forest Industries Association is pushing the government to develop criteria for certification. The Canadian Pulp and Paper Association [CPPA] has been instrumental in initiating the certification programme being developed by the Canadian Standards Association [CSA]. In other countries, where there has been little pressure to ban or boycott 'unsustainable' timber, neither industry nor the trade has been under pressure to think about certification.

Whatever their degree of support, much of industry and most traders are concerned about the possible increases in cost which certification – and by implication sustainable forest management – may lead to. They are concerned about the effects of this in terms of competitiveness with alternative products. They are equally concerned about the lack of clarity and transparency in current certification initiatives. They do not want to have to meet the multiple standards, and suffer the multiple audits, of multiple certification programmes, each established in different markets with little regard to one another.

Retailers and consumers

Retailer groups are the closest to the final consumer and the market. It is these groups who hope to reap the benefits from certification in the marketplace. Retailers are used to an environment where customer satisfaction is paramount. They expect certification to increase customer satisfaction and thereby to improve their business prospects. It is not surprising that retailers are, on the whole, strongly in favour of certification.

However, retailers tend to be against country-level certification. They recognize that forest practices differ within countries; and they are more used than other groups to dealing with individual suppliers. Indeed, many of their suppliers already work to detailed product specifications. This might include independent certification of particular product properties such as fire resistance or strength. It might also include a requirement for system certification such as ISO 9000. As a result, retailers often see certification as an additional product specification – an environmental specification which is part of a product's quality dimension.

Perhaps surprisingly, not all consumers are in favour of certification. There is still misunderstanding about the process and concern about its reliability.

However, most consumer groups are, by nature, 'pro-choice' and, where certification increases consumer choice in a reliable and non-misleading way, they support it. Given the visual way in which forest problems are often presented, these groups also tend to prefer programmes which reduce negative visual impacts, such as large clear-cuts.

Overall trends in the certification debate

The potential of certification is now generally accepted, although some groups still raise fundamental dilemmas which are summarized below. To many governments, trade groups and NGOs, certification now appears to be inevitable, and a better alternative to bans and boycotts. The hope is that certification could be a useful mechanism to reconcile the needs of free trade and economic, social and environmental sustainability. Few, however, see it as a solution in isolation. There is a general understanding that certification has to be part of a strategic approach to sustainable forest management: it is not a single-issue 'miracle cure'.

With better information and more debate, both the initial fears about certification and many of the initial naive hopes for it are being moderated. The standards debate had been holding many interest groups up, and the political ramifications, especially sovereignty issues, had been holding up government and intergovernmental consideration. These are now being eased through mutual learning and debate.

Interest in certification tends to be demand-led. Of governments, importing governments tend to be taking the lead, being more supportive of the idea than exporting governments, some of which still see certification as a possible trade barrier. However, this is changing. Similarly, timber importers are encouraging timber exporters to address the possibilities.

Some basic principles seem now to be agreed:

- Temperate, boreal and tropical forests all need to be included.
- Harmonization is needed for: acceptable standards of sustainability; and mutual recognition between different certification programmes.
- Standards and procedures should be set through wide stakeholder participation.
- Local interpretation of standards needs to be allowed for.
- Certification procedures and the accreditation of certification bodies should be clear and rigorous.
- Certification should be based on cooperation and transparency, not discrimination.
- Cost-minimization needs serious attention.
- The public needs to be educated about certification, thereby allowing bogus schemes to be exposed.

With this general agreement more or less in place, we are witnessing institutional positioning – in particular, of international institutions such as ITTO and ISO. Where there are gaps, new groups – notably the FSC – are being formed.

The sticky issues – differing views that need resolving

Most views seem to conclude, either reluctantly or enthusiastically, that the issue now is not whether certification should go ahead, but what should be the *responsibilities and mechanics* of certification – how to do it, how high to set the standards, how much regulation should complement the voluntary approach, what other complements are necessary, the costs and benefits for different groups, efficient procedures throughout the chain of custody, who should run it and who to involve. These 'sticky issues' should be the focus of research and fora for further debate. Continuing ITTO, CIFOR and FSC work will be helpful here. More legal study may be required. Many issues, however, cannot be resolved until there is more experience of certification on the ground. Support to existing initiatives and their evaluation is therefore desirable.

The effectiveness of certification in solving forest problems

The dilemmas are:

- whether certification offers an incentive to improve the management of the majority of production forests, as opposed to just rewarding the few forests that are already sustainably managed;
- whether certification will lead to systemic market changes, or establish only small 'niche' markets;
- whether certification will improve market access, or act as a trade barrier;
- whether certification alone is adequate, or whether complements such as life cycle analyses of forest products, or key regulations, are also required; and
- whether or not certification can be 'enforced', and the scope for illicit timber entering the market.

Who should run certification schemes

The dilemmas are:

- whether national schemes should be run by government, by industry, by NGOs or by partnerships; and
- whether international harmonization and accreditation should be run by an NGO or by a multilateral body such as the European Commission, and whether this should be in the UN system such as UNCTAD and GATT or outside the UN, such as ITTO. In contrast, NGOs tend to prefer independent – NGO-coordinated – watchdogs, considering verification and information credibility to be key.

The scope – all forests?

The dilemmas are:

- natural forests only; or including plantations, separately or integrated;
- the minimum size of forest to be certified; and
- whether or not conversion forests – ie those being converted from forest to non-forest cover – can be certified.

The standards and their assessment

The dilemmas are:

- the processes required for defining standards – principally the levels of stakeholder participation;
- the scope and detail of standards, especially regarding social issues;
- whether to go for minimum, achievable standards or to set maximum requirements;
- how to make assessment practicable in the forest; and
- the degree of local interpretation of standards that is desirable.

Phasing

The dilemmas are:

- whether to go for a slow approach led by pilots or by less stringent schemes, or to start with a major organizational and policy commitment covering rigorous standards. For example, the EU has suggested a stepwise approach beginning with country-level certification and moving then to the LFMU;
- whether to begin programmes with all forests, or whether tropical forest only would be best to start with – some importers favour the latter, because of domestic pressure to boycott tropical wood, eg Netherlands, Austria and Switzerland; and
- whether or not a period of grace should be included for compliance. Japanese industry prefers this; so do many ITTO Producer Member governments – who believe that it is only fair to begin certification after 2000, when ITTO's Objective 2000 should ensure a level 'playing field' of sustainably-managed forests.

Concessions for 'special cases'

The dilemmas are:

- whether or not special provisions need to be made for small producers such as community groups, to enable their forests to be certified; and
- whether or not support to poorer countries should be included as part of certification schemes; such as to cover the incremental costs required for them to establish and run certification programmes.

We hope that this handbook has assisted in providing material to help resolve these 'sticky issues'.

Conclusions on what to do next

In concluding the many ideas that have been presented in this handbook regarding certification, there are three levels of need which currently stand out:

- *At the forest level,* to ensure that both forest managers and certification bodies can accurately interpret external standards for specific sites. Moreover, to ensure that managers can implement [and certification bodies can assess] the management systems required to assure, over the long term, a level of environmental and social performance consistent with the certfication standards. Addressing this need must be possible for all LFMUs – regardless of size – and not prohibitively costly.
- *At the international level,* to agree one general international standard [probably expressed as principles and criteria]. This should be directly applicable at the site level and avoid certification being a barrier to trade [it should therefore allow for all forest types]. Creation of such an international standard requires improved information-sharing on certification; and greater understanding of the relationship between standards, accreditation and assessment - which are all inter-related within the certification process and should be considered together, not separately.
- *At the national level,* or regional if appropriate, to bring together stakeholder groups to explore the potential and implication of certification within the national forest context. An important need is to allow for government involvement in setting national guidelines for site interpretation of the international standard; and government involvement in accreditation of certification bodies. The potential for using existing [ISO and other] standard-setting procedures and national government accreditation bodies should be examined.

Annex 10.1 *Government viewpoint – Malaysian delegation to ITTC, May 1994*

The following is a summary of the intervention points raised by the delegation of Malaysia at the ITTO Working Party on Certification of All Timber and Timber Products held in Cartagena de Indias, Colombia from 12–14 May 1994. Our emphases are given in italics.

1 *Malaysia is not averse to timber certification.*
2 *In principle, Malaysia can appreciate the rationale that timber certification and labelling is being advanced in the name of sustainability.* However, there are such schemes being promoted in some consuming countries ostensibly in the name of sustainability but are governed actually by political and economic expediency. Some of these schemes are being developed unilaterally to cover [only] tropical timber.
3 *To ensure that certification, as a policy instrument, truly contributes to the attainment of sustainable forest management and access to markets, everything must be taken into consideration before any decision is taken regarding its implementation,* [so that it] will not give rise to new problems and complications.
4 *The purpose, role and limitation of timber certification and labelling in ensuring and enhancing the sustainability of the world's forests have to be clearly understood.* By and large, timber certification and labelling has so far been conceived as a trade policy instrument. Many studies have shown that trade in timber has not contributed significantly to deforestation and the volume of timber entering the international market is very much less than the total volume of global production. Underlying causes of forest degradation and destruction relate more to poverty, indebtedness, under-development, shifting cultivation and fuel needs.
5 *It is essential that certification applies to all types of timbers, while similar and comparable arrangements should be considered for timber substitutes including plastics, aluminium and steel.* Since the bulk of internationally traded timber is non-tropical, it makes sense to attach importance and priority to the certification and labelling of non-tropical timbers.
6 *Malaysia is concerned over the proliferation of unilateral proposals on timber certification and labelling.* There is a great deal of differing perceptions and interpretations regarding the issue of sustainability. First party efforts by producer countries are often viewed with scepticism and the need for verification has often been emphasized. Third party initiatives particularly in consumer countries have been advanced without adequate understanding of the actual situation in producer countries. The overall situation is quite confusing.
7 *Ideally, all existing unilateral efforts should be discontinued and replaced by a single global mechanism which applies to all timbers.* The mechanism should be agreed to by all relevant parties including governments, the timber trade, international organizations and non-governmental organizations from both producing and consuming countries.
8 *A credible and workable timber certification and labelling scheme will require a realistic and suitable time frame for implementation,* which shall not be sooner than the objective by which the sustainability of all types of forests is to be attained. In this regard, it appears that the year 2000 has gained increasing acceptance.
9 *There seems to be a clear correlation between the comprehensive and effective coverage of a certification scheme and its attendant costs.* The prospect for what is referred to as the green premium appears to be existent in some countries in Europe but not elsewhere.
10 *The issue regarding timber from conversion forests has to be addressed as it impinges on the right of countries to development.*

Source: Government of Malaysia delegation to ITTC, Cartagena, 1994

Annex 10.2 *The UK Government's views on timber certification and labelling*

Well designed and implemented certification and labelling schemes could offer useful market-based incentives to sustainable forest management, by enabling consumers to make choices between timber from sustainable and unsustainable sources. Certification and the monitoring that goes with it could also offer valuable practical benefits to timber producers through the provision of improved management information.

Much work still has to be done on the development of internationally-recognized, compatible and practically applicable standards, which are acceptable to a sufficiently wide range of governments, NGOs and end-users, for certification to be credible. Setting standards for forest management is primarily a matter for national governments, and certification and labelling needs to be acceptable to the governments of timber-producing countries. But schemes also have to be credible to individual timber purchasers, and it seems likely that many will seek non-governmental approval in order to ensure objectivity.

It is important that certification should reflect the multiple use of forests, and the fact that forest management is a dynamic process. It also should not unduly favour large-scale operators.

The Overseas Development Administration is helping the Soil Association to develop its Responsible Forestry Programme, particularly in developing countries. The Government, however, agrees that the potential benefits of certification and labelling schemes extend to temperate and boreal timber, which should equally be sustainably produced.

The Government does not believe that certification should be used by individual governments to impose import controls. Trade measures applied unilaterally by one country in order to secure particular ends in another would conflict with the UNCED Forest Principles, and would be contrary to our international trading objectives in the framework of GATT and the WTO.

Source: Ian Symons, *Overseas Development Administration* (1994) letter to IIED

Annex 10.3 *The Canadian Government's position on forest certification*

The federal government views the Canadian certification initiative positively and is supportive of it. Federal support is based largely on the belief that a voluntary sustainable forest management standard that is science-based and credible, in that it takes into account the interests of all the major forest stakeholders, will help to address concerns in Canada's international markets that purchases of Canadian forest products originate from companies practising sustainable forestry.

Regardless of the approach used to address the certification issue, there is a need to achieve a consensus among the various interest groups on the basis for a certification system, ie on the criteria that constitute sustainable forest management.

The Canadian forest products industry believes that this consensus can best be achieved by working through the independent processes employed by national and international standards-writing organizations. The industry believes that the Canadian Standards Association (CSA) should take the lead in the development of a national standard for sustainable forest management. The Canadian forest products industry is pursuing the development of a voluntary sustainable forest management systems standard. Development of this standard is being done by the CSA, which may ultimately lead to a global standard developed under the auspices of the ISO.

In June 1994, a Memorandum of Understanding (MOU) was signed between the CSA and the Canadian Pulp and Paper Association, which acts as the representative for the forest products industry on the issue. The MOU will result in the development of a guideline and specifications documents for voluntary Canadian standards for sustainable forestry certification. The inaugural CSA focus meeting was attended by a broad mix of forest stakeholders, including the federal and provincial governments, environmentalists, wildlife conservationists, the industry and others.

It is envisioned that a draft Canadian standard for sustainable forest management systems will be ready by Summer 1995, and finalized by December, 1995. An ISO standard, based on the CSA Canadian standard, could not be finalized until the end of 1998 at the earliest.

The Canadian industry believes that a sustainable forest management 'system' should form the basis of the national standard. Such an approach implies certification of an organization, rather than of its products, and means that firms would be evaluated/audited in terms of their ability to manage a forest site under their care in an environmentally sound and sustainable manner.

Evidence of sustainable forest management could be indicated by management process factors such as: forest policies, goals, objectives, planning auditing, records, staff training and responsibilities, public participation and regulatory compliance. Environmental performance indicators will likely be incorporated in the CSA standard. Being systems-oriented, the CSA sustainable forest management standard will be consistent with the ISO approach to EMS.

Canada is committed to strengthening global understanding of what sustainable development of forest means, and in particular, to addressing the question of how best to measure progress in achieving sustainable forest management. As a result of this commitment, the Department of Natural Resources Canada is actively involved in both domestic and international processes to develop criteria and indicators for sustainable forest management. The industry's initiative to develop a voluntary sustainable forest management standard and the development of criteria and indicators are linked. This linkage arises because a number of the criteria and indicators for sustainable forest management will likely form the basis for the environmental performance indicator component of the CSA sustainable forest management standard.

Source: Carette, J and Caron, S (1994) *Canadian Forest Service*, letter to IIED

Annex 10.4 *A view from the South*

Issues and challenges

1   There is no agreement on the organization of a certification system. The intended processes are still very confusing and non-transparent. Who has the right to certify? Who grants the mandate? Should certification be voluntary or mandatory? Should a unique 'official' certification alone be valid?

2   Some certifiers are operating from their own perspective and initiative, simply for gaining a market niche, though never having received a mandate to do so. *The governments of the developing countries are not participating at all in the process and they also have no defined role.*

3   None of the systems is using a macro approach to try to develop national certification procedures. They are using a case-by-case approach that focuses on private companies oriented to the export market. The existing micro approach will delay the progress of achieving the target of having certified sustainable forest management by the year 2000.

4   No initiative has been taken to finance and make certification accessible to every forest owner, including communities. And no parallel policy initiative has been taken to finance the higher costs of sustainable forest management. (Apparently there is much faith in the market mechanism, which until now has not favoured forest management.)

5   Certification should also be undertaken for products destined for domestic markets, if the objective is to stop deforestation and to have a global dimension. If a decision is taken in this direction, national governments should participate actively.

6   Certification groups are trying to apply *ideal* standards to forest management with too many principles, criteria and indicators, requiring too much information for the whole process to be practical. The result is that certification is not accessible to everybody and is very expensive. The result will be a low take-up of sustainable management principles and practices – only a limited adoption by those who can afford it, or who are obliged to become certified because of commercial constraints.

The certification process should therefore be reoriented as follows

1   Certification should *set minimum standards and not maximum requirements*. Since the problem is a global one requiring a global solution it seems logical that there should *only one type of certification which is general* and controlled by a higher body like the UNDP and/or ITTO.

2   All forest management units willing to be certified should receive *the same and only one category of certification*. Different categories only bring confusion to the consumers. We believe that forests *committed to* sustainable management should be certified, even though they may not yet have achieved all of the necessary minimum requirements. However, the certifier along with the owner must agree upon and determine a realistic timespan within which the owner must fulfil the remaining requirements for achieving sustainability in order to *maintain his certification*. In addition to being practical and realistic, this approach would also avoid the unrealistic and Utopian target

date of the year 2000 for achieving sustainability.

Certification is a dynamic process. Although only one category should be established, an agenda of progress towards complete sustainability would complement this. We first need to avoid non-sustainable practices and second to make progress towards sustainability. We don't want to set barriers, but instead to help countries, forest owners and communities to get on the road towards sustainability and to eventually achieve it.

Source: summarized from Ronnie de Camino, Cesar Sabogal, M and Martins, Paul J (1994) *Authorship and Expectations of Timber Certification Standards*

Annex 10.5 *View on the FSC from an environmental NGO: Pro Regenwald, Germany*

Fundamental flaws of the FSC and its Principles and Criteria (P&C)
The Forest Stewardship Council (FSC) aims to be 'an international body to accredit certifying organizations to guarantee the authenticity of their claims'. In an ideal world, such a scheme may be tenable. But this is not an ideal world, and the FSC initiative as it stands is inoperable. Not only does it have no legal basis, as a private body, but it takes no account of the realities of the illegal trade prevalent throughout the timber trading world, which would make monitoring virtually impossible. This and other fundamental flaws will ensure not only its failure but will severely hamper attempts to conserve the world's remaining natural forests.

The Principles and Criteria make no distinction between primary forests and other forests in terms of 'sustainable use'. There is no acknowledgement of the need to protect primary forests from commercial logging or the need actively to encourage non-destructive means of obtaining a living from these forests – 'non-lethal forest management'. There is no recognition of the Precautionary Principle in the 'sustainable' exploitation of all forests advocated by the FSC.

The accreditation scheme proposed by the FSC has no basis in either international or national law. With no legal basis to the scheme, there can be no serious attempt to monitor and ensure compliance, or initiate adequate enforceable measures to combat the illegal trade in timber.

Why the FSC will fail

1 It is trying to be everything to everybody: The P&C, despite a long consultative process, err on the side of the industry and the timber consumer, rather than conserving forests and enhancing the rights of indigenous peoples. The claimed aims of the FSC, 'to promote environmentally appropriate, socially beneficial and economically viable management of the world's forests', are mutually incompatible.

2 Market forces do not regulate trade: Experience shows that exploitation of natural resources based on international trade is invariably unsustainable. Market forces do not work to regulate voluntarily the trade they stimulate, the basic assumption of the proposed FSC accreditation scheme. Rather they tend to exhaust the resource, whether it be minerals, wildlife or timber.

3 Industry is too organized and laws are too weak: The FSC is up against an industry which is highly organized, with skilled public relations and marketing techniques. At least 99 per cent of trade is acknowledged as being unsustainable. Laws regulating it are often deficient and enforcement is weak or non-existent – conditions which strongly favour illegal trade. To try to implement a poorly-conceived, non-governmental worldwide accreditation scheme with no legal backing under such circumstances has no hope of success.

Why the FSC Principles constitute a charter for destruction

The principles proposed under the FSC initiative are not principles as such but a mixed bag of ideas that reflect the attempt that has been made to accommodate all constituencies. They are either completely impractical or they amount to nothing less than a charter for continued destruction – a 'Choppers Charter'. The FSC places far too much faith in the 'management plan'. Too little is known to produce management plans which are not in danger of further depleting existing forests. Far more research is needed into the ecological and social consequences of forest exploitation.

Conclusions and recommendations

The FSC is attempting the impossible. In a world in which our ignorance outweighs our knowledge, we cannot determine what is truly sustainable, least of all with respect to management of primary forests. Neither can we ensure that it will be complied with. *There is need to face the challenges of the present to overcome them, not to appease social and commercial practices that have brought us to where we are.* If, however, the FSC insists on continuing in its objective, a pilot project is all that can be attempted realistically at the moment. At the very least it could serve to highlight the myriad problems that will arise with any attempt to set up an accreditation scheme. But even to give a pilot project a chance of success requires a complete reassessment of the P&C, basing them not on 'sustainable' use of all forests but on a *precautionary approach* to forest management.

Source: Reeve, Dr Rosalind (1994) *The Forest Stewardship Council – Aims, Principles and Criteria: A critical examination predicting its failure*

Annex 10.6 *An environmental NGO's viewpoint: WWF-UK*

Background
Since 1990, WWF UK has been investigating the reliability of unverified claims of sustainable forest management and in 1991 produced a report of its findings, entitled *Truth or Trickery?* (Box 9.8) The report, which analysed many claims made for tropical timber products, concluded that the vast majority could not be verified.

WWF is working to see all these unverified claims removed from company literature, point-of-sale information and on-product labels, as we believe that at best they confuse, and at worst deliberately mislead, the consumer. We also believe that the Trade Descriptions Act (1968) should be amended to close the loophole which permits continuing self-certification and labelling.

Against this background it has been clear for some time that there is an urgent need for independent timber certification, to bring order to a confused marketplace.

The concept of an organization to evaluate, accredit and monitor wood and wood product certifiers, was first proposed by the Woodworker's Alliance for Rainforest Protection (WARP) and in 1991 the idea of the Forest Stewardship Council (FSC) was developed. Since then, WWF has been working with like-minded organizations and individuals:

- to *produce a working definition of 'sustainable forest management'* – the Forest Stewardship principles for natural forest management; and
- to *establish an international system for certifying 'well-managed forests'* and tracing the timber from them.

The next step: Support the FSC
A strong purchasing policy can stimulate a strong incentive for improved forest management, which in turn will ensure that suppliers conform to internationally-agreed standards of forest management. Companies and individuals who continue to buy cheap timber on the open market risk buying into the process of deforestation and forest degradation worldwide, some of which is being caused by the timber industry.

The FSC will take the guesswork out of deciding whether a particular shipment of timber has come from a well-managed forest. We urge companies to investigate how they can use the FSC and the certifying organizations to help them ensure their timber purchases come only from well-managed forests. WWF has been working with a number of companies which have been trying to do exactly that.

In conclusion
WWF calls on companies, local councils and architects and specifiers to adopt a timber product purchasing policy which:

- gives preference to suppliers who can provide independent evidence of good forest management;
- gives preference to 1995 group members;
- ignores timber labels and unverified certificates of sustainability; and
- requires that all wood product sources are certified by an independent body which has been accredited by the Forest Stewardship Council.

Source: Jeanrenaud, Jean-Paul and Sullivan, Francis (1994) *Timber certification and the Forest Stewardship Council: The WWF perspective*

Annex 10.7 *A certifier's viewpoint: Soil Association, UK*

Certification as a market-led incentive for bringing environmental improvements is very much in line with the ideology of our times. It is proving to be an important catalyst in bringing together the players who need to cooperate, in order to develop environmentally and socially sound forest management, to develop the mechanisms needed to verify good forest management, and to encourage new thinking and a change in attitudes.

However, certifiers can only *verify* the status of forests. They cannot themselves *create* sustainable forests. Indeed, the area of forest worldwide which could be certifiable is at present lamentably small. While certification currently benefits individual producers and traders who are already looking to the future, the goal of large-scale improvement of forest management must also be tackled at the political level, backed up by redeployment of international finance away from unsustainable forestry. At some stage, regulatory devices will need to be introduced if sustainable, certified timber is not to be marginalized into a 'niche' market, rather than becoming a perfectly normal commodity.

Past experience indicates that intergovernmental deliberations on forest issues will be slow and tortuous, with the very real risk of settling on the lowest common denominator that can be agreed by all parties. Nevertheless, one must remain hopeful that the various initiatives generated by the UNCED process will move international thinking forward, perhaps goaded by the examples of high quality forestry demonstrated by certification. Critical to these deliberations are transparency, participation of affected groups and accountability, if secure and meaningful agreements are to be reached.

While this laborious process unfolds, timber certifiers and others are developing standards for responsible forestry (crucial for consumer confidence), transparent, objective and relevant methods for assessing forest management, and chain-of-custody procedures; and are gathering information about the costs of sustainable forestry and market obstacles to trade in sustainable timber. All this is essential experience which must be acquired for a properly-regulated trade in sustainable timber products.

Source: Letter from Dorothy Jackson, Soil Association, to IIED (1994)

Annex 10.8 View of a forest peoples' group: The World Rainforest Movement

1   Certification must be part of a *regulatory framework*, not just a voluntary market initiative. The risk of a purely market approach is that certified timber will remain a niche product and certification will be unable to achieve its aim of eliminating destructive timber exploitation.
2   The real issue in certification is not the standards as laid down on paper, but the level of *control over those who are carrying out certification*. If certifiers' operations are not transparent, certification could easily legitimize the status quo.
3   The strong view of many Asian indigenous peoples and NGOs is that certification can only legitimize current practice because it will be *captured by current economic interests due to the politically repressive governments in these countries*. To allay these fears, it is essential that certification programmes are truly independent and open to challenge.

Source: Letter from Dr Marcus Colchester to IIED (1994)

Annex 10.9 *View from a conservation NGO: Soltrust (Solomon Islands)*

Soltrust believes that wood certification should be statutorily imposed. It should be the subject of a new world protocol, that should be debated and resolved upon by the UN if it is genuinely concerned about deforestation and degradation of the environment. This is a matter which we hope the FSC and others like WWF, Greenpeace etc should collectively begin to lobby for in the UN.

The proposed certification protocol should entail the enactment of national legislation by the signatory countries of the [certification] protocol to impose certification (in this regard Soltrust has already written to our government suggesting such legislation). It should also entail the introduction of preferential tariffs favouring certified wood and wood products. We do not believe this protocol would be an infringement on the new GATT agreement, because it should not discriminate between countries but between sustainable wood harvesting and non-sustainable wood harvesting.

Uncertified wood should be made illegal, or at the very least should be frowned upon in the same way as toxic waste, narcotic drugs and substances capable of nuclear manufacture are. Of course, forest plantations which are established for clearfelling should be exempted from this stringent protocol, because if this is the case we believe it would motivate most logging companies to start afforestation and reforestation activities. As we can appreciate, if this happens, it should reduce the need by industries to touch the natural rainforests and their ecosystems; this would be a plus for all of us.

Deforestation and environmental degradation has largely been due to the industrial activities of the now-rich developed countries, the G7 countries and the greed of its timber merchants. The governments of such countries should be asked to set up funds through levies they should impose for uncertified wood and wood products. These funds can be given as aid to Third World countries, especially those in the tropics, to re-establish natural rainforests and for logging companies to borrow from for forest plantation establishment.

Source: Letter from B Saua, Soltrust to IIED (1994)

Annex 10.10 *View of the Forest Industry Committee of Great Britain [FICGB]*

The consistent position of the forestry industry in Britain is that we are dedicated to sustainable forestry management, in conformity with the Statement of Forestry Principles agreed at UNCED 1992. Hence we support the principle of certification or an International Standard – provided that any procedure is practicable, does not prejudice the competitiveness of wood as against alternative products, and is undertaken by the proper accountable authority – which is the Forest Authority division of the Forestry Commission.

This means that any certification mechanism should be devised upon the basis of accountable, regulatory structures. Regulation is a legitimate function of Government. The integrity of the Forestry Commission, as the Government Agency accountable to Parliament for administering regulations and standards in British Forestry, is not in question.

Together with other countries, Britain is pursuing international negotiations begun after UNCED in 1992, aimed at achieving formal accord under the auspices of the UN Commission for Sustainable Development (UNCSD), which will set International

Current Initiatives and Views

Standards to be achieved by the national signatories and include the indicators and criteria against which forestry will be measured.

The Forest Stewardship Council (FSC) operates outside the aegis of the UN structure and does not recognize the principle of national sovereignty established by UNCED. While it promotes its 'independence' of government, the fundamental objection is that it is self-appointed, judgmental and non-accountable. It would not be acceptable for over 10 per cent of the British land area to become subject to a procedure where the arbitration or judgement of a self-appointed body could have a major and possible prejudicial effect on property and investment values, without any course of appeal. Within proper and legally-constituted procedures, by contrast, the law both enforces and protects.

Compliance with official regulations administered by an agency accountable to Parliament (and which amount to a certification procedure in all but name) cannot be subject to the risk of possible 'veto' (by withholding a 'certificate') by a non-accountable body, claiming the power to appoint inspectors or operating an alternative set of prescriptions which have no legal authority. While the 'independence' of the FSC might appeal to certain political views, this cannot be advanced as justification for usurping the official functions of government itself. The extension of the precedent to other economic, commercial or even social activities would point the way towards chaos or an environmental despotism.

The British Government, through the Scottish Minister of State, has already made two statements in answer to Parliamentary Questions that the Forestry Authority regulates the industry *'in accordance with the government policy to promote the sound management of woodlands as a renewable natural resource'* and that *'there is therefore no need for any other organization to carry out inspections or certification'.*

The British forestry industry has studied the proposals of various NGOs and conservation bodies which are seeking 'accreditation' by the FSC as nominated 'certifiers' in Britain. None have demonstrated the capability or knowledge of the industry to undertake such a role on the scale required in this country. Due to the lack of knowledge of the wood chain (and an apparent failure to research it), none of the procedures suggested are, in our view, workable. The potential costs are unacceptable and unjustified by such proposals, which merely seek to duplicate what are recognized to be among the most efficient regulatory structures in the world.

Source: FICGB presentation to the European Forestry Commission and ECE Timber Committee, Sweden (May 1994)

Annex 10.11 *A forest concession-holder's viewpoint: Alpi, Italy and Cameroon*

We are sensitive to the debate about good forest management, sustainability and certification which has arisen in the past few years.

The reasons for our commitment are twofold: first, we want to carry on our business in an ethically correct way. Second, being able to deliver ecologically safe tropical wood products is a powerful marketing tool in importing countries such as the US, Canada, Germany, Scandinavia, Netherlands, Austria and Switzerland.

Between 1990 and 1992 we hired a consultancy company in order to achieve a good level of low-impact forest management. We implemented their instructions and we think that we can claim today that our forestry operations are 'good'. We do not use the word 'sustainable' because, up until today, there does not seem to be a clear definition and consensus on this term.

Nevertheless, we have not been seeking certification yet for three main reasons.

1   The first problem is one of acceptability. The certifying company is being paid by the company that has to be certified. Various participants in the environmental debate (ie fundamentalist NGOs) are likely to dismiss certification on these grounds.
2   Second, certification costs should not be too high, and certification procedures should not be too cumbersome to make a project economically viable.
3   Third, the people operating in the field should have adequate preparation. In this regard, we have been working with several 'experts' who certainly had a good scientific background, but showed insufficient knowledge of commercial forestry practices. This kind of knowledge cannot be developed in a short period of time.

Source: Letter from Vittorio Alpi to IIED (1994)

Annex 10.12 *Viewpoint from a forest producer and paper manufacturer: Aracruz, Brazil*

Forest certification must:

1   be realistic;
2   be practical;
3   fit with other systems like quality assessment and environmental management;
4   have realistic costs;
5   be efficient;
6   aim at gradual improvement;
7   not be a trade barrier;
8   be a conflict-solver not a conflict-creator;
9   have a particular regional focus; and
10 not be anti-development.

Source: Letter from Claes Hall, Aracruz to IIED (1994)

Current Initiatives and Views

Annex 10.13 *View from the Netherlands Timber Trade Association*

The main issue in the debate is sustainable forest management, ensuring the continuity of the structure and functions of the forest, including sustainable timber production.

Given the principal causes of forest degradation in the tropical zone, the key element for improving the situation is large-scale socio-economic development, whereby the forest is to play its own important role.

Given the ownership of those forests, it is imperative that policy development in this respect, including certification schemes, should be pursued as joint actions with developing countries.

From our perspective, certification is only a vehicle to carry information on the sustainability of the source, from the production side towards the consumer and other interested parties.

Certification schemes should therefore be initiated preferably by, or at least in close cooperation with, timber-producing countries. Such schemes must be credible, internationally approved and ensure the sovereignty of the countries involved. They must be based on international criteria for sustainable forest management, which have been checked for their practicability in applying them in specific types of forest. Monitoring should be carried out by independent national or international bodies as a foolproof and credible basis for a certificate.

Without the active and voluntary involvement of producing countries, any initiative is bound to fail in its execution.

Source: Letter from Klaas Kuperus to IIED (1994)

Annex 10.14 *A retailer's viewpoint: Do It All, UK*

All timber, not just tropical

Seminars and debates are keenly addressing such issues as forest certification, wood tracing and labelling systems, and sustainable forestry. The arguments no longer deal with *whether* there is a need for these, but instead concentrate on *how* such schemes are to be set up and monitored.

The key word is credibility. We all have a role to play in designing and developing systems that are credible; and it is unfortunate that early environmental criticism focused on the tropical timber-producing countries, with almost no reference to the situation in temperate countries. There are countries, such as Germany and the Netherlands, which are embarking on schemes that smack of discrimination – the Netherlands 1995 Covenant and the German Projekt Tropenwald, which focus on tropical timbers, without reference to temperate forests. Such schemes threaten to derail significant improvements in forestry management and stop the establishment of credible schemes worldwide.

Forest certification must take place in all forests. There must be no double standards.

A united approach

Wood certification remains tricky: in any form, it will vary according to the country concerned, and it would be disingenuous to suggest that one system of wood certification can be used for all forests. But we are still learning. Environmentalists need to work with the industry; and the trade needs to adopt a more collaborative stance. It is inevitable that circumstances will change, largely owing to social, political, economic, ecological and silvicultural factors. One thing is for sure: we do need a network of some kind for the sake of consistency and credibility; and I am encouraged that the FSC has taken an important first step towards achieving this goal.

Paying for certification

There comes a time in every business activity when we have to incur costs to safeguard future interests. Can we allow synthetic materials to replace products from a renewable source? The cost of certification must be faced by the timber industry if it is to survive. How exactly that cost will be distributed throughout the supply chain is the subject of ongoing debate. The timber trade has been too passive for too long: it's time for it to take a more active role. You are either a part of the problem or a part of the solution. There is no middle way.

Going about it

It would be better for individual companies and associated forestry operations to voluntarily take up wood certification. Wood products which are certified will enjoy greater market access and share; and integrated systems which trace wood from the forest to the point of sale can also be used to provide precise quantity and quality information on wood movements through the supply chain.

A comprehensive wood certification scheme accompanied by a British Standard (BS)7750 environmental management programme, could bestow on participating companies a 'green' corporate image. Where a BS7750 and/or BS5750 programme is introduced alongside the certification process, companies may also see an increase in productivity rates.

Assessing a forest for certification purposes would begin by establishing the existence of a forest management plan and by checking that it conforms with the national forestry laws of the country in question. The management plan would then be compared with emerging international standards. Ironically, these procedures run parallel with quality assurance procedures as described in BS5750, ISO9000 and EN29000 – standards which many timber companies worldwide already comply with.

The challenge

The main obstacles relate largely to cost and to building up trust between the developed and developing nations. But what is the paying customer's role – and it's an increasingly important one – in all of this? As one managing director of a well-known timber company put it: 'Joe public is very good at telling industry one thing and doing something quite different with its money.' All the more reason for us to rewrite the *status quo!*

Source: Paul Ankrah, from articles in the *Timber Trades Journal* [1994]

Part 4

Directories

11. Active Certification Programmes

RAINFOREST ALLIANCE

Contact

Kate Heaton
Rainforest Alliance
65 Bleecker Street
New York
NY 10012-2420

Tel: +1 212 677 1900
Fax: +1 212 677 2187

Background

The Rainforest Alliance was established in 1987 as a non-profit organization whose aim is to work internationally to conserve tropical forests.

The Smart Wood Certification Programme began in 1989. This certifies well-managed and sustainable sources of tropical timber, as well as companies that sell certified wood in raw or finished forms in the US market. In 1994, Smart Wood expanded its scope of operations to include temperate and boreal forest regions in the US and Canada. In addition, the programme launched the Smart Wood Network, a cooperative effort to link regional certification organizations throughout the world.

Organization

Administration: Smart Wood is managed by a core staff of forestry specialists.

Evaluation: This is conducted by Smart Wood staff, advisors and consultants, together with a local specialist from the country involved. The composition of an assessment team is determined by the complexity of the forestry operation to be assessed.

Peer review panel: This consists of three independent specialists who examine the draft report submitted by Smart Wood staff.

Aims and Objectives

Smart Wood provides independent third-party evaluation of forests and companies, enabling consumers to identify products whose harvesting does not contribute to the destruction of forests. Smart Wood certifies wood or wood products that come only from 'sustainable' or 'well managed' forests.

By promoting the use of wood from well-managed forests, Smart Wood aims to encourage the adoption of sustainable forestry practices that meet long-term environmental, economic and social objectives.

Official Smart Wood policy is to develop publicly available and country specific Smart Wood certification guidelines for all countries where the Rainforest Alliance works.

Standards

Smart Wood generic forest management guidelines have been developed in consultation with professional foresters, ecologists and social scientists, and are revised continually. These criteria form the basis of field evaluations.

Forest assessments are based on three principles:

1   maintenance of environmental functions, including watershed stability and biological diversity;
2   sustained yield forestry production; and
3   positive impact on local communities.

Certified forests are classified according to how closely they adhere to these principles (Operational procedure, Step 6 below).

Where Smart Wood has developed country or region specific guidelines, these will supersede the generic guidelines.

Operational procedure

Step 1 – Submission of an application: All forest operations, or companies involved with selling wood and other forest products, are eligible to apply for certification. As much written documentation as possible should be submitted to ensure subsequent stages are carried out effectively.

Step 2 – Smart Wood response: This will occur within 30 days of application submission, after due consideration of all relevant issues and documentation.

Step 3 – Field assessment: The assessment team will:

*   review and revise the guidelines to incorporate country or region specific issues, and ensure coverage of relevant government legislation;
*   meet with government forestry specialists, environmental and community development non-government organizations and all other interested and affected parties;
*   visit field operations;
*   visit office operations to review the forest operation procedures and systems in place for maintaining detailed records; and

Directories

173

- conduct a final briefing with the field and office based staff to discuss future steps in the certification process.

Step 4– Peer review: The draft report will be submitted to a confidential peer review panel consisting of three independent specialists.

Step 5– Consultation: Smart Wood staff will consult with the forest operator to confer accuracy of all information.

Step 6– Certification decision: There are five possible options:

1　Certification as a 'sustainable' source: that which operates in *very strict* adherence to the Rainforest Alliance principles and guidelines.
2　Certification as a 'well-managed' source: that which can demonstrate a very strong *operational commitment* to Rainforest Alliance principles and guidelines.
3　Certification as one of the above but includes specific conditions that have been identified for improvement prior to the first annual audit.
4　Not to certify, with an explanation and stipulation of conditions that must be met in order to qualify in the future
5　Not to certify because there is not enough information; information gaps must be specified by Smart Wood and an agreement made to reconsider when the information has been provided.

Forestry operations can enrol in the Smart Wood programme as 'pre-certified members', but only after a field-level, pre-certification assessment has taken place. This type of programme enrolment is a formal indication that the enrolled organization has received recommendations for improvements, supports certification, and is working to convert its operation to full certification.

Smart Wood companies are certified according to whether all or some of their wood products come from certified Smart Wood sources: 'exclusive' and 'non-exclusive' Smart Wood companies.

All certified Smart Wood forests and companies appear on the periodically updated Smart Wood list.

Step 7 – Annual or random audits: Smart Wood requires at a minimum, annual audits, and reserves the right to conduct random audits.

SCIENTIFIC CERTIFICATION SYSTEMS

Contact

Debbie Hammel
Scientific Certification Systems
The Ordway Building
One Kaiser Plaza
Suite 901
Oakland
California 94612
USA

Tel: +1 510 832 1415
Fax: +1 510 832 0359

Background

Founded in 1984, Scientific Certification Systems (SCS) is a for-profit, multi-disciplinary scientific organization. Formerly known as Green Cross, it has developed a series of programmes that provide independent evaluation and certification for environmental and food safety performance.

SCS established its Forest Conservation Programme (FCP) in 1991.

Organization

FCP programme director: is responsible for implementing a FCP evaluation for a forest operation.

Evaluation team: consisting of contracted regional experts in the area of forestry, ecology and the social sciences.

Peer review team: consisting of entirely non-SCS personnel who must have recognized credibility and expertise in the geographic region of the evaluation. Their responsibilities are to review case-specific final reports and the overall FCP evaluation protocols and criteria.

Aims and objectives

The FCP is a third-party forest management certification programme designed to distinguish and recognize 'well-managed' forest operations in which timber products are produced in a manner that sustains the timber resource, maintains the forest ecosystem, and meets minimum financial and socio-economic criteria.

It provides uniform guidelines for assessing the relative sustainability of forestry operations worldwide and an independently verified basis for potential market-place claims.

It aims to encourage incremental improvements in forest management practices by:

- creating internal management auditing systems;
- providing baseline data on existing management practices;
- clarifying and facilitating the development of long-term forest management goals;
- monitoring the movement towards long-term goals through periodic audits; and
- providing marketplace recognition through a labelling and promotional system.

Standards

Before an evaluation starts, the evaluation team decides upon a working set of evaluation criteria for each of the three programme elements described below. These criteria are selected from a pre-established set of base criteria and are judged to be most relevant to the evaluation.

Programme Element 1 – Timber resource sustainability: Covers the way in which the forest is managed for continuous production over time. Evaluation criteria include:

- harvest regulation;
- stocking and growth control;
- pest and pathogen management strategy;
- forest access;
- harvest efficiency and product utilization; and
- management plan and information base.

Programme Element 2 – Forest ecosystem maintenance: Covers the extent to which the natural forest ecosystem is adversely altered during the process of managing the forest and extracting timber products. Evaluation criteria include:

- forest community structure and composition;
- long-term ecological productivity;
- wildlife management actions, strategies and programmes;
- watercourse management policies and programmes; and
- ecosystem reserve policies.

Programme Element 3 – Financial and socio-economic considerations: Addresses the financial viability of the forest operation and its socio-economic impacts. Evaluation criteria include:

- financial stability;
- community and public involvement;
- public use management;
- investment of capital and personnel; and
- employee and contractor relations.

The elements and criteria described above relate specifically to natural forest management, but similar elements and criteria have now been developed for the evaluation of plantation forests.

Operational procedure

Forest assessment:

Step 1 – Initial meeting: to clarify that the forest operation's goals are compatible with those of FCP.

Step 2 – Preliminary evaluation: to ascertain the probability for a successful certification, the cost and expected time frame. A brief written report is submitted upon completion.

Step 3 – Execute the contract: the signing of an agreement between SCS and the forest operation.

Step 4 – Assemble the evaluation team: under the direction of the Team Leader this team is responsible for steps 5–8.

Step 5 – Collect and analyse data.

Step 6 – Select and weight criteria: for the three programme elements.

Step 7 – Assign numerical performance scores: using importance-weighted criteria, an assessment is undertaken to determine the extent to which the forest management meets the underlying objectives associated with each evaluation criterion. The final result is a score (0–100) for each programme element. If the importance-weighted aggregate scores for each of the three programme elements exceed 60 points, the operation ownership qualifies for certification as 'well-managed'.

Step 8 – Technical report: written by the evaluation team, this is submitted to the client for feedback before it is sent on to the peer review committee.

Step 9 – Peer review: the reviewers will comment on both the general FCP methodology as well as the results of a specific evaluation.

Directories

Certification is valid for a time period expressly stipulated by SCS and is usually coincident with the landowner's management planning cycle. Certified forest operations are required to have an annual on-site audit by SCS in order to maintain their certified status. In addition to annual audits, SCS reserves the right to conduct irregularly-timed short-notice inspections.

Chain of custody procedures

For those companies wishing to identify certified products through product labelling, SCS has a list of requirements that can be drawn upon for implementing case-specific chain of custody procedures for any point of product transfer. Annual on-site audits are required to maintain a 'certified' chain of custody status.

Labelling

SCS allows for the following:

- Specific point-of-purchase claims: can be made by companies that have had either all their sources certified or only a portion but have implemented chain of custody procedures to ensure proper segregation.
- General non-point-of-purchase claims: can be made regarding certification of specific forest areas. This is recommended for companies who are unable to implement chain of custody procedures, and consequently must specify what percentage of their timber supply comes from certified sources.
- General point-of-purchase claims: are not affixed to a product but appear as a banner and can be made by a company without chain of custody procedures.

SGS FORESTRY

Contact

Frank Miller
SGS Forestry
Oxford Centre for Innovation
Mill Street
Oxford
OX2 0JX
United Kingdom

Tel: +44 (0)1865 202345
Fax: +44 (0)1865 790441

Background

SGS Forestry is part of the international SGS group of companies which specializes in inspection and quality certification. SGS Forestry has offices in over 140 countries worldwide and 32 000 employees.

SGS Forestry's certification programme was started on 1 April 1992.

Organization

Certification council: consisting of a minimum of five senior professionals representing forest industry, environmental and social interests, its main purpose is to ensure that the documented certification procedures are adequately implemented, and that assessments are carried out in an impartial and professional manner. It also adjudicates over all appeals.

Programme manager: is responsible for the day to day running and implementation of the certification programme.

Assessment staff: will carry out the forest assessment according to tried and tested procedures to ensure objectivity and consistency. Forest assessors have generally attended independently-run assessor courses, and the quality of their work is regularly monitored.

Aims and Objectives

The SGS Forestry certification programme covers all forests. It aims to promote quality forest management that meets the level of performance required by the Forest Stewardship Council Principles and Criteria (FSC P&C). Quality forestry is defined as that which is:

- *environmentally sensitive:* when its impact upon the environment is both assessed and minimized.
- *socially beneficial:* when it recognizes its activities have an impact both on local people and society at large.
- *economically viable:* when forest operations are structured and managed in such a manner that they are sufficiently profitable so as to ensure the stability of operations.

SGS Forestry's certification procedures and approach has been built around those used in Quality Management Systems standards (such as ISO 9000) and Environmental Management Systems standards (such as BS 7750). Importance is placed on a well-structured forest management system that will monitor and identify potential problems well before they become a significant concern, rather than an approach based on the setting of prescriptive levels for various management outputs.

Standards

Instead of setting its own standards, SGS Forestry has an assessment system which is based on clearly documented procedures. This is described in an Assessors Handbook which specifies the issues, described below, which must be taken into consideration during an assessment. The handbook provides the structure by which the assessor can judge whether or not forest operations are performing at the level required by the FSC P&C.

- Compliance with legislation, agreements and treaties, including:
  - policy
  - register of legislation
  - communication mechanisms
  - policing and review.

- Social elements, including:
  - policy
  - tenure and use rights
  - indigenous peoples' rights
  - employment rights
  - local participation.

- Benefits from the forest, including:
  - optimal economic use
  - waste reduction
  - alternative uses of forest resource (social and economic).

- Environmental impact of forest management operations, including:
  - policy and planning
  - impact assessment
  - objectives and targets.

- Forest management systems, including:
  - management plan
  - operator training
  - monitoring of performance
  - mechanisms for progressive improvement in performance
  - use of plantations.

Operational procedure

Step 1 – Scoping visit: this is a pre-assessment visit to determine if the current management system and operation is sufficiently developed to justify a formal assessment. It also enables the assessor to gain a clear appreciation of how the component parts of the whole production chain fit together, and to schedule and cost the full assessment. A standard checklist is used to ensure all issues are covered. If significant gaps in the management process and operations are apparent at this stage, then the assessment process is stopped until these are rectified.

Step 2 – Document review: depending on the size of the operation, some or all of the following documentation is required for examination:

- A public-stated policy: covering environment, social aspects etc.
- A published set of standards to which the company operates.
- A management plan encompassing all national legislation that might apply to forest operations.
- A plan of operations consisting of simple written practical procedures describing how all practical and administrative operations are to be carried out.

The aim is to ensure that relevant legislation, as well as actual standards, are addressed in a systematic fashion. In this way their implementation can be guaranteed over an extended period.

Step 3 – Site audit: This typically involves a team of three, comprising a lead forest auditor, a quality assessor and a regional independent expert who has detailed local forestry knowledge. The team verifies that the documented management system is being implemented to a level that conforms with the standard.

Step 4 – Peer review: Three experts from the panel review the formal report, which contains the findings of the assessment team, and provide feedback on whether

Directories

they believe the report presents an objective, balanced and technically sound assessment.

Step 5 – Certification and registration: Formal award of the certificate and entry of the forest details into the Directory of Registered Organizations.

Step 6 – Periodic surveillance and review: Routine visits to assessed organizations on an annual or 6-monthly basis, to ensure that the original levels of performance are being maintained and improvements being made. If major problems are identified then SGS Forestry reserves the right to suspend the certificate, until such time as the problem has been rectified. In extreme circumstances the certificate may be withdrawn.

THE SOIL ASSOCIATION

Contact

Ian Rowland
Soil Association
86 Colston Street
Bristol, BS1 5BB
United Kingdom

Tel: +44 (0)117 9290661
Fax: +44 (0)117 9252504

WOODMARK

THE SOIL ASSOCIATION'S
CERTIFICATE OF
RESPONSIBLE FORESTRY

Directories

Background

The Soil Association is a registered charity and was founded in 1946.

In 1992 the Soil Association established its Responsible Forestry Programme to carry out education and research into responsible timber production and to provide an ethical and well-regulated basis for the certification of timber. This was to be achieved through the timber certification and labelling programme (Woodmark).

Following two years of development, during which time standards were drawn up and inspection and certification procedures tested, Woodmark became operational. By mid-1994 two British woodlands had been certified under the Woodmark name.

Organization

Board of the Soil Association's marketing company: This is responsible for directing the timber certification programme.

Advisory board: This consists of practising foresters, forestry academics, ecologists, social scientists and representatives of the timber trade, this advises on technical and policy matters.

Inspections: These are carried out by Woodmark staff or on a contract basis by qualified professionals. Local inspectors are used where possible.

Administration: This is carried out by four staff members: a coordinator, two forest officers and an administrator. Volunteer assistance is also available.

Certification committee: This consists of professionals with expertise appropriate to the certification, this is responsible for reviewing inspectors' reports.

Aims and objectives

Woodmark aims to link the producers of timber and timber products from responsibly-managed forests with consumers who wish to buy such products. In this way, the implementation of good forestry practices will be encouraged. In addition, the aim is to improve company efficiency, information gathering, record keeping and quality control at management and production levels.

Where appropriate, Woodmark will work closely with local counterparts and organizations, to cooperate in building the foundations of national certification organizations.

Woodmark will operate under the auspices of the Forest Stewardship Council (FSC). Until the FSC accreditation programme is established, Woodmark will operate its own system of approval for other certification schemes or certified forests. This will allow forests and forest products certified by other schemes to be formally recognized, thus avoiding duplication of certification activity.

Standards

Woodmark is based on the following 6 generic principles:

Principle 1 – Environmental impact: Forest management minimises negative impacts on the biodiversity, soils, water and landscapes of the forest and adjacent areas.

Principle 2 – Sustained yield: Yields of forest products and services are sustainable in the long-term.

Principle 3 – Land rights: Legal land rights of indigenous and traditional peoples are established and enforced. Customary use rights to the forest are maintained.

Principle 4 – Local control, consent and benefit: Indigenous and traditional communities control forestry activities on their lands. Forestry operations receive the full and informed consent of local communities, enhance their long-term social and economic well-being and do not reduce their ability to make use of the forest in any way.

Principle 5 – Economic potential: Forest management encourages an optimal and efficient use of all forest products and services, in order to ensure a wide range of environmental, social and economic benefits.

Principle 6 – Management and monitoring: A written plan appropriate to the scale and resources of the forestry operations clearly defines the ownership of the forest and describes the ecology of the forest. It states the objectives of management and how they will be achieved and explains how the long-term maintenance of forest products, services and social benefits is assured, and specifies monitoring procedures.

The standards cover forest management and chain of custody as well as setting out criteria for operations, application, registration and control over the use of the Woodmark label.

These standards are generic in that they deal with general forest management worldwide, and are not specific to place or forest type. They apply to natural and modified forests, plantations and timber and timber products. They define the minimum management standards which all forests worldwide should meet.

Under Woodmark, additional local forestry standards can be drawn up. These address specific silvicultural, social and economic conditions in different countries or geographic regions. This has been achieved for forestry in the UK. Adaption of the generic standards in this way, takes place through consultation with relevant forestry authorities and all interested groups.

Operational procedures

Step 1 – Application: A completed application form, accompanied by supporting documentation, is screened by certification staff. A desk study or preliminary visit may be required to obtain further information prior to setting up the inspection.

Step 2 – Inspection: Ground inspections are carried out to determine how closely forest operations confer to the Woodmark's principles of responsible forestry.

Step 3 – Review: The certification committee meets to review inspectors' reports. It recommends whether the operation should be certified, based on current standards of operation, and the scope for further improvement.

Step 4 – Certification: Conditional certification will be awarded where minor irregularities exist, on the understanding that there will be demonstrable improvements or provision of further information within a defined time frame. Successful forest operators sign a contract agreeing to abide by the certification regulations, and are licensed to use the Woodmark label and wording on specified products. Licenses are renewed annually following successful reinspection and payment of the annual licensing fee.

Directories

FOREST STEWARDSHIP COUNCIL

Contact

Dr Timothy Synnott, Executive Director
Forest Stewardship Council
Avenida Hidalgo 502
68000 Oaxaca
Oaxaca
Mexico

Tel:+52 951 62110
Fax: +52 951 62110

Background
(See also Box 9.5)

The Forest Stewardship Council (FSC) was founded in Toronto, Canada in October 1993. The Founding Assembly was attended by 130 participants from 25 countries, who voted to legally constitute the FSC as an independent, non-profit, non-governmental membership organisation.
   Since then the FSC has:

* ratified nine of its ten Principles and their accompanying Criteria (P&C);
* established its headquarters in Mexico;
* appointed an executive director; and
* established an international board of directors who meet quarterly.

Organization

General Assembly

This is the highest authority in the FSC. It comprises two chambers:

1 That which represents economic interests and has 25 per cent of the vote. Membership can be sought by any organization or individual with a commercial interest in the forest industry so long as they have demonstrated active commitment to implementing the FSC's P&C in their operations.
2 That which represents social and environmental interests and has 75 per cent of the vote. Membership is restricted to non-profit, non-governmental organizations, indigenous organizations and social movements committed to environmentally appropriate, socially beneficial and economically viable forest management.

Board of directors

- Voted in by the General Assembly, this is structured so as to achieve a balance between social, ecological and economic interests as well as North and South.
- Board members must be demonstrably committed to the FSC. Certification Bodies and industry associations may not be represented on the Board.

Secretariat

Headed by the executive director, this executes the decisions of the Board and runs the day-to-day operations of the organization.

Technical Committee

This reviews and makes recommendations to the Board on such matters as the development of FSC P&C and national and regional standards.

Accreditation Committee

This reviews the application of Certification Bodies and proposes the conditions of acceptance to the Board.

Dispute Resolution Committee

This deals with disputes and grievances which members may have concerning the performance of the executive director, the secretariat or the Board.

Regional representatives

With the establishment of several regional offices, the FSC aims to decentralize its work and encourage local participation.

Aims and objectives

The FSC aims to establish worldwide standards for good forest management. The FSC will complement this aim by means of the promotion of the P&C, the accreditation of Certification Bodies and by supporting the regional development of local standards based on the P&C.

Mission Statement

The FSC will promote management of the world's forests that is:

- *environmentally appropriate:* such that the harvesting of timber and non-timber products maintains the forest's biodiversity, productivity, and ecological processes;
- *socially beneficial:* such that both local people and society at large enjoy long-term benefits from harvesting forests;
- *economically viable:* such that forests are properly valued, the prices of forest products reflect the full costs and benefits of good management, and sufficient reinvestment is made in the forest resource.

Coverage

The P&C apply to all tropical, temperate and boreal forests which are managed for production of forest products. The intention is that the P&C be regarded as a super-set of standards from which more focused national or regional standards are developed. The P&C also apply to plantations.

Accreditation procedure

Certification Bodies can apply voluntarily to the FSC for accreditation and the right to use the FSC name in their labels. The FSC will assess these requests based on the Certification Body's adherence both to the global FSC P&C and evidence to support the existence and use of adequate assessment procedures.

Documentation

Certification Bodies must submit a formal application to the FSC accompanied by the following documentation:

- *Letter of Intent:* expressing the Certification Body's interest and intent to become accredited.
- *Organization and infrastructure:* of the Certification Body.
- *Certification procedures:* details of the Certification Body's procedures and methods for evaluating and certifying forest management enterprises.
- *Chain-of-custody procedures:* details of the Certification Body's procedures and methods for monitoring the distribution of forest products.
- *Operations:* list of forest areas where the Certification Body has carried out assessments, where a certificate was awarded, and countries and forest operations for which the Certification Body wishes to be accredited.

If the application is accepted, the FSC and the Certification Body will enter into a contract for evaluating whether accreditation can be granted.

Evaluation

The evaluation team will examine the following aspects:

- *Forest Management Standards:* standards and procedures used by the Certification Body for evaluating forest management must adhere to the P&C and to any existing FSC-recognized national standards of forest management.
- *FSC certification guidelines:* the Certification Body's system for evaluations, and monitoring must adhere to the FSC Guidelines for Certifiers. These include:
  - compliance with the FSC P&C;
  - independence from government or timber industry influence;
  - sound evaluation procedures;
  - complete transparency and openness to scrutiny by the FSC;
  - reciprocity; it is expected that certificates issued by an accredited Certification Body are mutually recognized by other accredited Certification Bodies;
  - making appropriate information about certification activities available to the public;
  - documented procedures for verifying the chain of custody;
  - compliance with applicable laws;
  - maintenance of a fair and non-discriminatory evaluation process;
  - maintaining adequate documentation;
  - appeal procedures;
  - maintenance of proper control over the use of licences, logos, certification marks and the name of certifier.
- *International and national standards organization:* the Certification Body's methodology must adhere to relevant standards of ISO and national and regional standards organizations.

Accreditation

The decision to accredit the Certification Body is made ultimately by the Accreditation Committee. Accreditations will be for a stipulated period not exceeding five years, at which time a re-evaluation must be conducted .

At least biannually, FSC will conduct a monitoring inspection of an accredited Certification Body's operations and records to assure continued compliance with the conditions of accreditation. The FSC reserves the right to conduct short-announcement inspections of an accredited Certification Body's operations and records.

Directories

FSC Principles and Criteria

The FSC's Principles and Criteria (P&C) apply to all tropical, temperate and boreal forests. Many of these P&C apply also to plantations and partially replanted forests. More detailed standards for these and other vegetation types may be prepared at national and local levels. The P&C are a complete package to be considered as a whole, and their sequence does not represent an ordering of priority. Each principle is accompanied by several criteria.

Principle 1: Compliance with Laws and FSC Principles
Forest management shall respect all applicable laws of the country in which they occur, and international treaties and agreements to which the country is a signatory, and comply with all FSC P&C.

Principle 2: Tenure and Use Rights and Responsibilities
Long-term tenure and use rights to the land and forest resources shall be clearly defined, documented and legally established.

Principle 3: Indigenous Peoples' Rights
The legal and customary rights of indigenous peoples to own, use and manage their lands, territories and resources shall be recognized and respected.

Principle 4: Community Relations and Workers' Rights
Forest management operations shall maintain or enhance the long-term social and economic well-being of forest workers and local communities.

Principle 5: Benefits from the Forest
Forest management operations shall encourage the efficient use of the forest's multiple products and services to ensure economic viability and a wide range of environmental and social benefits.

Principle 6: Environmental Impact
Forest management shall conserve biological diversity and its associated values, water resources, soils, and unique and fragile ecosystems and landscapes, and by so doing, maintain the ecological functions and the integrity of the forest.

Principle 7: Management Plan
A management plan – appropriate to the scale and intensity of the operations – shall be written, implemented and kept up to date. The long term objectives of management, and the means of achieving them, shall be clearly stated.

Principle 8: Monitoring and Assessment
Monitoring shall be conducted – appropriate to the scale and intensity of forest management – to assess the condition of the forest, yields of forest products, chain of custody, management activities and their social and environmental impacts.

Principle 9: Maintenance of Natural Forests
Primary forests, well-developed secondary forests and sites of major environmental, social or cultural significance shall be conserved. Such areas shall not be replaced by tree plantations or other land uses.

Principle 10: Plantations
Plantations should be designed and managed consistent with Principles 1–8 and certain stated criteria. Such plantations can and should complement overall ecosystem health, provide community benefits, and provide a valuable contribution to the world's demands for forest products.

12. International and National Certification Initiatives

AUSTRIA

Contact
Joseph Hackl
Umweltbundesamt
(Federal Environmental Agency)
Spittelauer Lande 5
A – 1090 Wien
Tel: +43 1 31304 ext 314
Fax: +43 1 31304-400

Background
In 1990 the Austrian Parliament passed a resolution to stop the import of all tropical timber and tropical timber products from countries which do not practice sustainable forestry, according to accepted criteria.

Consequently in June 1992 a Federal Law was established 'for the labelling of tropical timber and tropical timber products and for the creation of a quality mark for timber and timber products from sustainable forest management'. The law provided for the mandatory labelling of all tropical timber and timber products being placed on the market.

In 1993 however, the Austrian Parliament amended this law after it was challenged on the grounds that this type of labelling was discriminatory, unjustifiable and an unnecessary obstacle to trade. Although mandatory labelling requirements were dropped, the voluntary quality mark still remains.

Certification initiative
Following the amendment of this law, an Advisory Board was established in 1993/1994. It was set up to consider suitable standards for defining sustainable forest management and to write a proposal for a labelling scheme. The Board consists of representatives from governmental organizations, the timber industry, social and economic interest groups and environmental non-governmental organizations.

Current situation/future position
Selected standards will be ready for testing during the summer of 1995 possibly in cooperation with the Centre for International Forestry Research (CIFOR). The standards will apply to all forests and aim to be compatible with existing international standards so as to acquire universal acceptance.

Following their establishment, efforts will be directed towards making the standards operational.

BRAZIL

Contact
Rubens Garlipp
Sociedade Brasileira de Silvicultura
Rua Marselha
1.180 – Bairro Jaguaré
São Paulo
SP-CEP 05332-000
Tel: +55 11 869 4941
Fax: +55 11 869 0798

Background
Since 1992, the Brazilian forestry sector has been developing a methodology for a programme of certification that will define the origin of raw materials used by the forest industry of Brazil.

This work has resulted in CERFLOR (Certificate of Origin of Forest Raw Material).

Certification initiative
The CERFLOR regulations are based on the assumptions of self-regulation, transparency, adaptation to Brazilian conditions, non-discrimination, voluntary application, flexibility and compatibility with international standards.

There are five CERFLOR principles, designed to cover planted and natural forests:

1   care for biodiversity;
2   sustainability of natural resources and their rational use in the short and long term;
3   care for water, soil and air;
4   environmental, economic and social development of areas of forestry activities; and
5   compliance with governmental legislation.

The principles and criteria have been developed to take account of regional forestry variations.

All the principles and criteria will be subjected to an impartial evaluation, although they do not all have to be fulfilled at the time of certification. CERFLOR will grant the certificate only if all its five principles have been complied with.

Current situation/future position
CERFLOR has developed a set of criteria and indicators for plantations and a methodology for evaluation and the subsequent award of a 'CERFLOR seal'.

Once both of these have been field tested and finalized, specific indicators will be developed and tested for natural forests.

Directories

CANADA

Contact
Jean-Claude Mercier
Université Laval
Quebec G1K7P4
Tel:+1 418 656 2131 ext. 3858
Fax:+1 418 656 2809

Background
The Canadian forest products industry declared its support for developing and implementing a certification programme for Canadian forests. The rationale was that only by demonstrating sound stewardship would Canada be able to maintain its competitive position in the international marketplace.

Thus in 1993, the Canadian forest industry set up a task force to review the need for certification and recommend the means for its development.

Certification initiative
The recommended certification programme incorporates a technical standard and an accreditation organization that accredits certifying bodies, which are authorized to issue certificates after independent, third-party audits of forest operations.

In order for the standards to be credible, practical in application and internationally acceptable, they should be externally set and represent broad stakeholder concerns about environmental impacts on forest management. Importance was placed on establishing a single set of standards for Canada, as it was perceived that regional or sub-sectorial approaches would only cause confusion in the international marketplace.

Current situation
The CSA (Canadian Standards Association) has formed a Technical Committee that is responsible for the development of sustainable forest management standards. These standards are based on ISO 9000 and 14000 standards and reflect environmental, economic and social values.

As a result the Technical Committee has developed two standards (see Box 9.7): *Z808 standard:* the SFM (Sustainable Forest Management) requirement and framework for planning, implementing, certifying and improving a SFM system; and *Z809 standard:* the SFM systems specification which specifies the auditing procedures related to principles outlined in the Z808 document.

These standards are expected to be finalized and available by June 1995, consequently allowing for the implementation of sustainable forestry certification by the end of 1995.

Directories

FINLAND

Contact
Sari Sahlberg
Ministry of Agriculture and Forestry
Department of Forest Policy
PO Box 232
FIN-00171 Helsinki
Tel: +358 (0)15621
Fax: +358 (0)1562232

Background
In June 1994 the Council of State declared that the implementation of sustainable forest management would be provided with the necessary policy environment including regulations and economic steering mechanisms.

Following this, in July 1994, the Ministry of Agriculture and Forestry and the Ministry of the Environment published a new environmental programme for forestry in Finland. The aim is that the programme will be continually reviewed and updated to take into account developments in sustainable forestry. These two developments form the basis of the Finnish forest certification initiative.

Certification initiative
By the end of 1994 preliminary studies had been completed by four working groups to explain existing forestry standards, listing those that were relevant but currently missing. Following this, a formal consultative process was launched by the Ministry of Agriculture and Forestry in December 1994. These discussions are focusing on appropriate standards for conservation and enhancement of biological diversity in forest ecosystems, and on the maintenance of broad socio-economic functions.

Current situation/future position
The framework for national standards is scheduled to be completed by the end of 1995.

The aim is for the standards to be scientifically competent, technically feasible and measurable. It is thought that assistance with the setting of procedures for forestry inspection, certification, supervision, complaint handling etc will be obtained from ISO (International Standardization Organization) guides and European standards.

The decision to establish a national certification programme has not yet been finalized. The process will certainly be a slow one as it will involve achieving and maintaining open discussion and consensus between forestry and environmental organizations. Importance is placed on opportunities for developing Nordic (at least between Finland, Sweden and Norway) cooperation in the certification process.

Directories

195

GERMANY

Contact
Stefan Schardt
Initiative Tropenwald (ITW)
Am Kollnischen Park 2
10179 Berlin
Tel: +49 30 2790132
Fax: +49 30 2793728

Background
Timber certification discussions began in December 1990 and pressure was put on the German timber industry to reach agreement on the use of tropical timber.

Thus in February 1992, a joint declaration was made on the protection of tropical forests, by the German Union for Furniture, Timber, Plastics and Allied Trades (GHK), Central Federation of German Timber (HDH) and German Timber Importers Federation (VDH). This demanded that only timber from sustainable sources should be imported and processed, and hence timber certification was required.

Certification initiative
In summer 1992, ITW was founded by the timber workers union, the timber processing industry and timber importers.

The ITW philosophy is based on the principles of cooperation not confrontation; incentives not sanctions; practical support not wishful thinking; and renewable raw materials where *possible*, not where *necessary*.

The certification system should be non-discriminatory, voluntary and non-governmental, although recognized by governments.

It is intended that ITW will contribute to the international discussion on developing a European certification institution which will monitor and control country of origin certificates on the European market.

Current situation/future plans
The Tropenwald Committee is still working on a certification programme, but has already established the following:

- a definition of sustainability;
- very detailed criteria and indicators for the evaluation of sustainable forest management. These are being tested by CIFOR;
- chain of custody procedures;
- requirements and structure of a certification body, according to EN 45011.

ITW have drawn up a strategy for the realization of a European Timber Certification System.

INDONESIA

Contact
Emil Salim
CPIS
Jalan Merdeka
Selatan No.13 Jakarta Pusat 10110
Tel: +62 21 344 6007/ 344 6008
Fax: +62 21 380 6210

Background
In September 1993, a preliminary set of standards was completed by the working group of experts set up by concession-holder and forest industry organisations (MPI/APHI).

This was followed by the establishment, in December 1993, of a Foundation for Eco-labelling (Lambaga Ekolabel Indonesia, LEI) under the chairmanship of Professor Emil Salim.

Certification Initiative
LEI is responsible for the development and operation of the certification and labelling scheme for Indonesian forest products, based on internationally accepted standards and processes.

In order to retain credibility, LEI will adhere to principles of independence and transparency. The standards applied will be based on a range of sources, including the Forest Stewardship Council's Principles and Criteria. Three sub-groups have been formed to cover ecological, economic and social forestry issues. To achieve international acceptance the selection and accreditation procedures for certification bodies will follow EN 45011.

LEI will accredit and commission certification bodies to undertake assessments and monitor their work. LEI will also review the audit reports and make the final decision regarding certification of the audited forest operation.

Under the LEI programme chain of custody verifications will be undertaken by the Ministry of Forestry.

Current situation/future plans
Following field testing of the forestry standards in June 1994, a manual is being written explaining them as well as their application.

It is aimed to have the LEI programme, including chain of custody verification, operational by the year 2000, in line with the ITTO target. Training and initial installation should already be underway by 1996.

THE NETHERLANDS

Contact
Eric Wakker
Milieudefensie
(Friends of the Earth Netherlands)
Damrak 26
1012 LJ Amsterdam
Tel: +31 20 6 221 366
Fax: +31 20 6 275 287

Background
In 1993 a legally binding framework agreement was signed by the government, timber trade representatives, labour unions and environmental NGOs. This is referred to as the Covenant on Tropical Timber or 'Netherlands' Framework Agreement on Tropical Timber'. All the signatories agreed to strive for the purchase of only sustainably produced tropical timber by the year 1995.

Due to conflicting attitudes on wood originating from conversion forests, the environmental NGOs have since decided to opt out of the framework agreement.

Certification initiative
A special commission has developed a framework designed to arrive at a definition of sustainable forest management and the government has taken the initiative to start dialogues with the main tropical countries exporting wood to the Dutch market.

During 1994, the partners of the framework committed themselves to take the necessary measures to make a certification programme operational. It was decided that when certified timber enters the country a national certification programme needs to endorse and harmonize the various certificates.

Current/future position
There is now a clear distinction in the Netherlands between the Covenant partners, now limited to the trade, the government and the labour unions, and a new Tropical Timber Campaign run by environmental NGOs. This is referred to as the HART VOOR HOUT ('Heart for Wood') initiative and is run by Friends of the Earth Netherlands and Novib.

Due to the lack of action since the signing of the Covenant, the NGO initiative itself is now preparing to promote certification. It aims to have non-sustainably produced tropical timbers removed from the market as soon as possible through government (EC) regulation. HART VOOR HOUT is actively promoting the Forest Stewardship Council.

The Dutch government has recently announced that it intends to postpone its 1995 target and link up with the ITTO year 2000 objective.

Directories

SWEDEN

Contact
Mr Anders Lindhe
Ulriksdals Slott
S-170 71 Solna
Sweden
Tel:+46 (0)8624 7435
Fax:+46 (0)885 1329

Background
Sweden's forest industry has been involved in discussions on sustainable forestry during the last few years, which has highlighted gaps in Swedish forest policy. In particular debate has focused on biodiversity and the need to gain more knowledge and incorporate this into forest practices.

The aim is to produce a formalised code of practice which will incorporate the necessary changes to forestry practice.

Certification initiative
Since 1993 representatives from all interested parties, primarily environmental and economic, have been involved in the process of drawing up a draft form of national sustainable forest management standards for Sweden.

This is a WWF-Sweden initiative which the Sweden Society for Nature Conservation recently joined. The process has focused on the interpretation of the Forest Stewardship Council's Principles and Criteria (FSC P&C) so that they will be applicable to Swedish forestry.

Current situation
Based on the FSC P&C, criteria for biodiversity are being discussed and a public draft was launched at the WWF Seminar 'Forests for Life', in May 1995.

It is hoped that a working organization will be established by the end of 1995.

Directories

SWITZERLAND

Contact
Pierre Hauselmann
CH 1523 Granges-près-Marnand
Tel/ Fax: +41 (0)37 64 20 87

Background
Between September and December 1992 the FSC undertook a national consultative process in Switzerland. This coincided with growing public concern on the sustainability of the world's forests.

Following this, the Swiss government declared that there would be a Swiss tropical wood label before the end of the year and appointed SGS Forestry to undertake a study as to how this could be achieved.

Certification initiative
The study was completed in 1993 and presented a proposal for a Swiss Certification programme which comprised the following elements:

- standards should be harmonized with those appropriate for other countries;
- separate provision for domestic and imported wood;
- a foundation to be established to accredit inspection bodies, to act as a licensor and promoter of the label;
- independent accredited certification bodies would carry out domestic forest auditing, labelling of wood products from audited Swiss forests, and verification of certification claims for imported wood;
- close cooperation with FSC.

At the beginning of 1994, a working group was established called Noyau FSC Switzerland. The group is made up of members from all interested parties including the forest industry, social development organizations, and conservation and environmental NGOs.This group commissioned a detailed feasibility study for the creation of a national FSC body in Switzerland. The study outlined the structure of such an organization and produced a first version of appropriate standards for Swiss forests.

For the remainder of 1994, Noyau FSC Switzerland evaluated the proposals made and considered how best chain of custody inspections should take place in Switzerland.

Current situation/future plans
Deliberations on the first part of the feasibility study have been completed and the second part is now underway. This intends to finalize and test the initial proposals.

Appropriate standards for Swiss forests are being tested in the Canton of Solothurn, and trial chain of custody inspections are planned in cooperation with DIY and furniture stores and with cantonal authorities in Basel-Stadt and Geneva.

It is expected that FSC Switzerland will be formally established by the end of 1995.

13. Certified Forests

This is a complete list of independently-certified forests as of June 1995

AMACOL LTDA, PORTEL, PARA, BRAZIL

Certified by: RAINFOREST ALLIANCE
Contact: Larson Wood Products Inc
31421 Coburg Bottom Loop
Eugene, Oregon, 97401 USA
Tel: + (1) 503 343 5229   Fax: + (1) 503 343 3279

Date of certification
1 November 1991

Duration
5 years

**Chain of custody inspection
     undertaken by**
Rainforest Alliance

Area of forest certified
59 000 ha

Volume produced per year
25 000 m^3

Main species/products
Veneer
Kapok-tree (*Ceiba pentandra*)
Faveira (*Vatairea spp*)
Breu sucuriuba (*Protium spp*)
Esponja (*Parkia spp*)
Plywood/pre-conditioned platforms
Assacu
Possumwood (*Hura crepitans*)
Muratinga (*Brosimum*)
Caju acu (*Anacardium giganteum*)
and other species

BAININGS COMMUNITY-BASED ECOFORESTRY PROJECT, RABAUL, PAPUA NEW GUINEA

Certified by: SGS FORESTRY
Contact: Max Henderson
Pacific Heritage Foundation
PO Box 546 Rabaul, Papua New Guinea

Tel: + (675) 96 4866   Fax: + (675) 96 4867

Date of certification
July 1994

Duration
5 years

Area of forest certified
12 500 ha

Volume produced per year
1200 m^3

Main species/products
Mixed 'Redwood' Mouldings –
dominant species is Taun

BROADLEAF FOREST DEVELOPMENT PROJECT, HONDURAS

Certified by: RAINFOREST ALLIANCE
Contact: Director
Proyecto Desarrollo del Bosque Latifoliado
PO Box 427
La Ceiba, Honduras

Tel: + (1) 504 43 1032   Fax: + (1) 504 43 1032

Date of certification
February 1991

Duration
5 years

Chain of custody inspection undertaken by
None

Area of forest certified
25 000 ha

Volume produced per year
1650 m^3

Main species/products
Cumbillo (*Terminalia amazonia*)
Sangre real (*Virola koschnyi*)
Cedrillo (*Huertea cubensis*)
Rosita (*Hieronima alchornioides*)
Laurel negro (*Cordia megalantha*)
and other species

CHINDWELL DOORS, JOHOR, MALAYSIA

Certified by: SGS FORESTRY
Contact: Dr Lincoln Chin
Chindwell Company Ltd
Hyde House The Hyde
London, NW9 6JT

Tel: + (44) (0)181 205 6171   Fax: + (44) (0)181 205 8800

Date of certification
July 1994

Duration
12 months initially

Chain of custody inspection undertaken by
SGS Forestry

Area of forest certified
3284 ha

Volume produced per year
7500 m^3

Main species/products
Rubberwood doors

COLLINS PINE, CHESTER, CALIFORNIA, USA

Certified by: SCIENTIFIC CERTIFICATION SYSTEMS
Contact: 1618 SW First Avenue
Suite 300
Portland
Oregon, 97201 USA

Tel: + (1) 503 227 1219   Fax: + (1) 503 227 5349

Date of certification
26 March 1993

Area of forest certified
38 300 ha

Duration
5 years

Volume produced per year
100 230 m^3

Chain of custody inspection undertaken by
Some inspections: SGS Forestry

Main species/products
Hemlock Fir
Ponderosa Pine
Sugar Pine
Douglas Fir
Incense Cedar

COLLINS PINE, KANE HARDWOOD, UNITED STATES

Certified by: SCIENTIFIC CERTIFICATION SYSTEMS
Contact: Wade Mosby
1618 SW 1st Ave Suite 300
Portland
Oregon 97201 USA

Tel: + (1) 503 227 1219   Fax: + (1) 503 227 5349

Date of certification
25 October 1994

Area of forest certified
48 300 ha

Duration
5 years

Volume produced per year
Lumber 13 344 m^3
Veneeer: 1420 m^3

Chain of custody inspection undertaken by
None

Main species/products
Veneer
Cherry
Red Oak
White Oak
Lumber
Cherry
Red Oak
Soft Maple
Hard Maple

DARTINGTON HOME WOOD, DEVON, UNITED KINGDOM

Certified by: SOIL ASSOCIATION
Contact: Silvanus Services Ltd
15 Link House
Leat Street
Tiverton
Devon, EX16 5LG

Tel: + (44) (0) 1884 257344

Date of certification
July 1994

Duration
12 months

Chain of custody inspection
undertaken by
None

Area of forest certified
92 ha

Volume produced per year
293 m^3

Main species/products
Douglas Fir
Larch
Spruce
Thuja
Sequoia

DEMERARA TIMBER LTD, GUYANA

Certified by: SGS FORESTRY
Contact: Janet Croucher
Lot 1 Water Street and Battery Road
Kingston
Georgetown, Guyana

Tel: + (592) 2 53835   Fax: + (592) 2 71663

Date of certification
June 1994

Duration
12 months initially

Chain of custody inspection
undertaken by
SGS Forestry

Area of forest certified
500 000 ha

Volume produced per year
39 000 m^3

Main species/products
Greenheart
Purpleheart
Morabukea/Mora
Demerara Mahogany
Baromalli
Locust
Red Cedar

KEWEENAW LAND ASSOCIATION LTD, UNITED STATES

Certified by: RAINFOREST ALLIANCE
Contact: Paul Tweiten
E 5090 Jackson Rd.
Ironwood
MI 49938 USA

Fax: + (1) 906 932 5823

Date of certification
1 November 1994

Duration
3 years

**Chain of custody inspection
   undertaken by**
None

Area of forest certified
50 000 ha

Volume produced per year
32 000 m^3

Main species/products
Hard Maple
Red Maple
Basswood
Yellow Birch
and other hard and soft woods.

MENOMINEE TRIBAL ENTERPRISES, UNITED STATES

Certified by: SCIENTIFIC CERTIFICATION SYSTEMS
Contact: Director
PO Box 680
Keshena
Wisconsin, 54135 USA

Tel: + (1) 715 799 3896

Date of certification
14 February 1992

Duration
5 years

**Chain of custody inspection
   undertaken by**
None

Area of forest certified
97 500 ha

Volume produced per year
28 500 m^3

Main species/products
Sugar Maple
E White Pine
E Hemlock
American Basswood
Red Maple
Yellow Birch
and other species

PENGELLI FOREST, DYFED, WALES – UK

Certified by: SOIL ASSOCIATION
Contact: Dyfed Wildlife Trust
7 Market Street
Haverfordwest
Dyfed, SA61 1NF

Tel: + (44) (0) 1437 765462

Date of certification
May 1994

Duration
1 year

Chain of custody inspection
undertaken by
Soil Association

Area of forest certified
65 ha

Volume produced per year
No volumes available

Main species/products
Oak
Ash – being made into chairs

PERUM PERHUTANI, JAVA, INDONESIA

Certified by: RAINFOREST ALLIANCE
Contact: Lynn-Nusantara Marketing Co Inc
21 East 28th Avenue
Suite D, Eugene
Oregon, 97405 USA

Tel: + (1) 503 686 9886   Fax: + (1) 503 485 6846

Date of certification
August 1990

Duration
5 years

Chain of custody inspection
undertaken by
Rainforest Alliance

Area of forest certified
2 831 500 ha

Volume produced per year
730 000 m^3

Main species/products
Teak
Mahogany
Rosewood
Pine

PLAN PILOTO, MEXICO

Certified by: RAINFOREST ALLIANCE
Contact: Plan Estatal Forestal
Henning Flachsenberg
Infiernillo 157
Esqu. Efrain Aguilar
Chetumal
Quintana Roo, Mexico

Tel: + (52) 983 24424   Fax: + (52) 983 24424

Date of certification
31 August 1991

Duration
6 years

**Chain of custody inspection
   undertaken by**
None

Area of forest certified
95 000ha

Volume produced per year
11 000 m^3 actually cut
28 000 m^3 potentially authorized

Main species/products
Amapola (*Pseudobombax ellipticum*)
K'atalox (*Swartzia cubensis*)
Bari (*Calophyllum brasiliensis*)
Chechen (*Metopium brownei*)
and other species

PLAN PILOTO FORESTAL, MEXICO

Certified by: SCIENTIFIC CERTIFICATION SYSTEMS
Contact: PIQRO
Bloohm Floors Inc
777 Brickell Avenue
Suite 1010
Miami
Florida, 33131 USA

Tel: + (1) 305 381 8070   Fax: + (1) 305 379 9509

Date of certification
26 January 1991

Duration
5 years

**Chain of custody inspection
   undertaken by**
None

Area of forest certified
33 000 ha

Volume produced per year
Unknown

Main species/products
Honduran Mahogany
Spanish Cedar
Ramon

PORTICO S.A., COSTA RICA

Certified by: SCIENTIFIC CERTIFICATION SYSTEMS
Contact: Royal Mahogany Products Inc
6145 – I Northbelt Parkway
Norcross
Georgia, 30017 USA

Tel: + (1) 404 729 1600   Fax: + (1) 404 446 8884

Date of certification
16 February 1993

Duration
5 years

Chain of custody inspection undertaken by
None

Area of forest certified
3900 ha

Volume produced per year
Unknown

Main species/products
Doors made from Carapa (Royal Mahogany) and Gavilan

SEVEN ISLANDS LAND MANAGEMENT COMPANY, UNITED STATES

Certified by: SCIENTIFIC CERTIFICATION SYSTEMS
Contact: 304 Handcock Street
Suite 2A
PO Box 1168
Bangor
Maine, 04402-1168 USA

Tel: +(1) 207 947 0541

Date of certification
8 November 1993

Duration
5 years

Chain of custody inspection undertaken by
None

Area of forest certified
406 250 ha

Volume produced per year
196 000 m^3

Main species/products
Black Spruce
White Spruce
Red Spruce
Balsam Fir
N White Cedar
Sugar Maple
Yellow Birch
E White Pine
and other species

TROPICAL AMERICAN TREE FARMS, COSTA RICA

Certified by: RAINFOREST ALLIANCE
Contact: TATF
717 City Park Avenue
Columbus
Ohio, 43206 USA

Tel: + (1) 614 443 5300   Fax: + (1) 614 444 0160

Date of certification
1 May 1994

Duration
3 years

**Chain of custody inspection
   undertaken by**
None

Area of forest certified
1336 ha

Volume produced per year
Lumber not yet available

Main species/products
Peroba Rosa (*Aspidosperma megalocarpon*)
Brazilian Rosewood (*Dalbergia nigra*)
Madero Negro (*Gliricidia sepium*)
Teak (*Tectona grandis*)
Nargusta (*Terminalia amazonia*)
and other species

Bibliography

AFPA (1995) *Sustainable Forestry Principles and Implementation Guidelines*, American Forest and Paper Association, USA

Anon (1994a) 'The Helsinki Process' *Ministerial Conference on the Protection of Forests in Europe: European Criteria and Indicators for Sustainable Forest Management*, July 1994

Anon (1994b) *Tropical Forest Update*, 4, (5) December 1994: International Tropical Timber Organisation, Yokohama, Japan

Anon (1995a) *The Montreal Process: Criteria and Indicators for the Conservation and Sustainable Management of Temperate and Boreal Forests*, Canadian Forest Service, Quebec, Canada

Anon (1995b) *Preliminary Criteria for Environmental Certification of Swedish Forestry 1995*, Swedish Society for Nature Conservation and World Wide Fund for Nature, Sweden

Baharuddin haji Ghazali and Simula, M (1994) *Certification schemes for all timber and timber products*, ITTO, Yokohama, Japan

BSI (1994) *Specification for environmental management systems*, BS:7750:1994, British Standards Institute, UK

Carew-Reid, J; Prescott-Allen, R; Bass, S; and Dalal-Clayton, D B (1994) *Strategies for National Sustainable Development: a Handbook for their Planning and Implementation*, Earthscan, London

CCFM (1995) *Criteria and Indicators of Sustainable Forest Management: The Canadian Approach*, Canadian Council of Forest Ministers, Canada

CSA (1995) *Sustainable Forest Management Systems – Specification Document Z809*, Working Draft 002, Canadian Standards Association, February 15th 1995

FAO (1995a) 'Expert Meeting on Harmonisation of Criteria and Indicators for Sustainable Forest Management', *Background Note 1/95*, FAO, Rome

FAO (1995b) 'Expert Meeting on Harmonisation of Criteria and Indicators for Sustainable Forest Management', *Background Note 2/95*, FAO, Rome

FSC (1994) *Forest Stewardship Principles and Criteria for Natural Forest Management*, Forest Stewardship Council, Oaxaca, Mexico

Grayson, A J (1994) *The World's Forests: Initiatives Since Rio*, Commonwealth Forestry Association, Oxford, UK

Groves, M A and Lambert, A J (1995) 'Environmental management systems and forest certification: a discussion of merits', *Eco-management and Auditing*, 2, pp 18–23, London

Heaton, K (1994) *Perspectives on Certification from the Smart Wood Certification Program*, Rainforest Alliance, New York, USA

Heuveldop, J (1994) *Assessment of Sustainable Tropical Forest Management: A contribution to the development of concept and procedure*, Mitteilungen der Bundesforschungsanstalt für Forst- und Holzwirtschaft, Hamburg, Germany

IIED and WCMC (1994) *Forest Resource Accounting: Stock-Taking for Sustainable Forest Management*, IIED, London

ISO (1994a) *Environmental Management Systems – Specification with guidance for use*, Draft BS ISO 14001 Document 94/400 681, British Standards Institute, UK

ISO (1994b) *Environmental Management Systems – General Guidelines on principles, systems and supporting techniques*, Draft BS ISO 14000, Document 94/4000 682, British Standards Institute, UK

ITTO (1990) *ITTO Guidelines for the Sustainable Management of Natural Tropical Forests*, Yokohama, Japan

ITTO (1991) *ITTO Guidelines for the Establishment and Sustainable Management of Planted Tropical Forests*, Yokohama, Japan

ITTO (1992) *Criteria for the Measurement of Sustainable Tropical Forest Management*, ITTO, Yokohama, Japan

ITTO (1993) *ITTO Guidelines on the Conservation of Biological Diversity in Tropical Production Forests*, ITTO, Yokohama, Japan

IUCN/UNEP/WWF (1991) *Caring for the Earth*, IUCN, Gland, Switzerland

IUCN/UNEP/WWF (1980) *The World Conservation Strategy*, IUCN, Gland, Switzerland

de Klemm, C (1993) *Biological Diversity Conservation and the Law: Legal Mechanisms for Conserving Species and Ecosystems*, IUCN Environmental Policy and Law Paper No 29, The World Conservation Union (IUCN), Gland, Switzerland

Official Journal of the European Communities (1992) *Council Regulation (EEC) No. 880/92 of 23 March 1992 on a Community Eco-Label Award Scheme*, 11 April 1992, No. L 99, 1, Brussels

Palmer, J (1995) *First CIFOR Test of Forest Stewardship Standards*, Forstamt Bovenden, Lower Saxony, Germany

Poore, D (1989) *No Timber Without Trees*, Earthscan, London

Pretty, J N and Howes, R (1993) *Sustainable Agriculture in Britain: Recent Achievements and New Policy Challenges*, IIED, London

Rainforest Alliance (1993) *Generic Guidelines for Assessing Natural Forest Management*, revised draft, Rainforest Alliance, New York, USA

Scientific Certification Systems (1994) *The Forest Conservation Programme: Programme Description and 1994 Operations Manual*, Scientific Certification Systems, California, USA

SGS Forestry (1994) *Assessors Handbook, Policy Document and Procedures Manual*, SGS Forestry, Oxford, UK (unpublished)

Speechly, H (1994) 'Standards and Forest Certification', *paper presented at the Canadian Pulp and Paper Association (Woodlands Section) 79th Annual Meeting*, Montreal, Canada

The Soil Association (1994) *Responsible Forestry Programme: Responsible Forestry Standards*, The Soil Association, Bristol, UK

World Commission of Environment and Development (1987) *Our Common Future*, WCED, New York

WWF (1994a) *Truth or Trickery? Timber labelling past and future*, WWF, Godalming, UK

WWF (1994b) *Criteria of Forest Quality: A submission to the Commission on Sustainable Development (CSD)*, WWF, Godalming, UK

Glossary

Definitions follow standard ISO/IEC terms where these exist.

Accreditation Authority – an independent third party which examines the organizational structure, responsibilities, practices, procedures, processes and resources of a Certification Body in the pursuit of certification activities.

Accreditation Schedule – a document issued by the Accreditation Authority which details those areas which a Certification Body has been assessed as competent to operate within.

Applicant [for certification] – person or body that seeks to obtain a licence from a certification body.

Assessment – Third party examination and evaluation of an LFMU's management systems to determine the degree of conformity against a specified standard; and to validate their effective implementation.

Assessment Schedule – a document issued by a certification body which is linked to a certificate and which details those areas for and within which a registered organization has been assessed as conforming with stated criteria.

Assessor – A person who is qualified and is authorized to perform all or any portion of a certification assessment.

Assessor's Handbook – A document which contains guidance and information to assist an assessor in the performance of his or her assessment activities.

Best Forest Management Practices – Practices which are at least equivalent to quality forestry and which are recognised as the best currently achieved for a particular forest type by an LFMU.

Biodiversity – A measure of the richness of life forms, often divided into ecosystem, species and genetic diversity.

Certificate of Conformity – document issued under the rules of a certification system, indicating that adequate confidence is provided that a duly identified product, process or service is in conformity with a specific standard or other normative document.

Certification of Conformity – action by a third party, demonstrating that adequate confidence is provided that a duly identified product, process or service is in conformity with a specific standard or other normative document.

Certification Body – body that conducts certification of conformity.

Certification Programme – a system that has its own rules of procedure and management for carrying out certification of conformity.

Chain of Custody – The monitoring process of the production and distribution channel from forest to end-product.

Community – all the people living in or adjacent to a forest in self-defined groups.

Continual improvement – year-on-year enhancement of overall environmental performance, not necessarily in all areas of activity, resulting from continuous efforts to improve in line with the LFMU environmental policy.

Corrective Action Requests [CARs] – Formal document which is raised by an assessor during assessment or surveillance which details non-compliances identified and remedial measures required within a specified time frame.

Criteria – key elements that define Principles.

Directory of Registered Organizations – Formal documentation usually published annually by a certification body which details those organizations which currently hold certification awarded by that body. It may also detail information provided on the related Assessment Schedules.

Environment – the forest surroundings and conditions in which an LFMU operates, including living systems (human and other). As certain environmental effects of the LFMU may reach all parts of the world, the environment in this context extends from within the LFMU to the global system.

Environmental Effect – Any direct or indirect impingement of the activities, products and services of the LFMU upon the environment, whether adverse or beneficial.

Environmental Effects Evaluation – A documented evaluation of the environmental significance of the effects of the LFMU's activities, products and services.

Environmental Effects Register – A list of the significant environmental effects, known or suspected, of the activities, products and services of the organization.

Environmental Management – Those aspects of the overall management function, including planning and the FMP, that determine and implement the Policy Document.

Environmental Management Audit – A systematic evaluation to determine whether or not environmental objectives and targets have been met.

Environmental Management Review – A formal evaluation of the status and adequacy of the LFMU's environmental policy, systems and procedures in relation to environmental issues, regulations and changing circumstances.

Environmental Management System [EMS] – The organizational structure, responsibilities, practices, processes and resources for implementing Environmental Management.

Environmental Objectives – The broad goals arising from the environmental policy and effects evaluation that an organization sets itself to achieve, and which are quantified wherever possible.

Environmental Policy – A public statement of the intentions and principles of action of the organization regarding its environmental effects, giving rise to its objectives and targets.

Environmental Targets – Detailed performance requirements, quantified wherever practicable, applicable to the organization or parts thereof, that arise from the environmental objectives and that need to be set and met in order to achieve those objectives.

Forest Management Plan [FMP] – Basc document adopted to describe how the LFMU will be managed.

Forest Manager – Person principally responsible for the activities of the LFMU.

Governing Body – The body which oversees the assessment and certification activities of the certification body, having equal and impartial representation of interested parties.

Guidelines – Overall directives which encompass and generally describe Principles and Criteria.

Indicators – See Environmental Effects.

Inspection Body [for certification] – body that performs inspection services on behalf of a certification body.

Interested Parties – See Stakeholders

International Organisation for Standardization [ISO] – world-wide federation of national standards bodies which produces international agreements published as international standards.

Lead Assessor – an assessor who is qualified and is authorized to manage a certification assessment.

Licence [for certification] – document issued under the rules of a certification system by which a certification body grants to a person or body the right to use certificates or marks of conformity for its products, processes or services in accordance with the rules of the relevant certification scheme.

Licensee [for certification] – person or body to which a certification body has granted a licence.

Local Forest Management Unit (LFMU) – the forest area to which the certification process applies.

Mark of Conformity [for certification] – protected mark, applied or issued under the rules of a certification programme, indicating that adequate confidence is provided that the relevant product, process or service is in conformity with a specific standard or other normative document.

Market-based Instrument – an economic instrument that affects costs and benefits of alternative actions open to forest managers, commercial and industrial stakeholders, with the effect of influencing behaviour in a way that is favourable to the forest environment.

National Forest Service – national government forestry organization responsible for regulation of activities in the forestry sector.

Natural Forest – any forest or closed woodland resulting from natural regeneration consisting almost entirely of native species.

Non Timber Forest Products (NTFPs) – All goods and services produced from the forest covered by the LFMU other than wood.

Peer Review Panel – the independent panel set up and monitored by the Governing Body to examine and comment upon assessment reports prior to final certification and registration of applicant organizations.

Permanent Forest Estate [PFE] – the area of forest in a particular country designated and demarcated as permanent; and usually divided into production and protection forest and subcategories of these two groups.

Plantation Forest – an area planted with trees of one or more species, usually but not exclusively for wood production.

Policy Document – first-level documentation of the formal internal management system operated by the LFMU. This document states the policy of forest management to which the LFMU is committed. It includes an Environmental Policy. It can also include other policy commitments such as a quality policy and confidentiality policy.

Precautionary Principle – action to prevent possible incidences of environmental or other damage, without waiting for scientific proof of the possible cause or effect of such damage.

Principles – Key elements of a Standard which define its scope.

Procedures Manual – second-level documentation of the formal internal management system operated by the LFMU. This document contains the operating procedures of the LFMU – including the FMP.

Production Forest – the area of natural and plantation forest within the PFE that is designated to be managed for productive purposes in perpetuity.

Quality Forestry [QF] – forestry which is environmentally sensitive, socially aware and economically viable (refer to Chapter 1 for a more detailed definition).

Reference Levels – See Targets.

Registered Organization – An LFMU which has undergone satisfactorily the process of assessment and certification and which has been formally entered on the Directory of Registered Organizations.

Stakeholders – Individuals and organizations with a legitimate interest in the goods and services provided by an LFMU; and those with an interest in the environmental and social effects of an LFMU's activities, products and services. They include those exercising statutory environmental control over the LFMU, local people, employees, investors and insurers, customers and consumers, environmental interest groups and the general public.

Standard – documented agreements containing technical specifications or other precise criteria to be used consistently as rules, guidelines or definitions of characteristics, to ensure that materials, products, processes and services are fit for their purpose.

Stewardship – management of forests with due regard for the long-term and external effects of activities; assuming responsibility for actions; accepting accountability; and adopting principles of good husbandry.

Supplier – the party that is responsible for the product, process or service and is able to ensure that quality assurance is exercised. The definition may apply to manufacturers, distributors, importers, assemblers, service organizations and so on.

Surveillance – Periodic re-examination and validation [see Assessment].

Tenure – Socially-defined agreements held by individuals or groups (either recognized by law or customary norms) on the rights of access and the rules for use of either a land area or of associated resources, such as individual trees, plant species, water, or minerals.

Use Rights – Rights for the use of forest resources that can be defined by local custom, mutual agreements, or prescribed by other entities holding access rights. These rights may restrict the use of particular resources to specific levels of consumption or particular harvesting techniques.

Other forestry titles from Earthscan

Forest Politics: The Evolution of International Cooperation
By David Humphreys

'*an important and timely book*' from the Foreword by Stanley Johnson

Global deforestation and its attendant processes – including soil degradation, climate change and the loss of biological diversity – emerged as international political issues during the 1980s, prompting politicians to seek consensus on programmes and policies for the conservation and sustainable management of forests. Yet global initiatives have been bedevilled by tensions between the North and the South and between governments, industry, local communities and indigenous peoples. Meanwhile, rates of deforestation in the tropics are increasing, and international political efforts are demonstrably failing.

Forest Politics carefully traces the evolution of international cooperation on forests, from the inception of the controversial International Tropical Timber Organization and the failed Tropical Forestry Action Programme in the mid-1980s, to the creation of the Intergovernmental Panel on Forests in the mid-1980s. The book also provides a detailed analysis of the negotiating stances of the parties involved in the divisive negotiations that took place prior to the 1992 'Earth Summit' in Rio de Janeiro and the equally factious negotiations for the International Tropical Timber Agreement of 1994. It provides a fascinating insight into the nature of such processes, illustrating the difficulties that arise when concepts such as 'global commons' come into conflict with national sovereignty.

Complete with annexes of important political documents, and making extensive use of primary source material and interviews with participants, *Forest Politics* presents case studies of all the major forestry negotiations over the last 13 years. It is an essential reference point for policy makers, environmental campaigners and students, and required reading for all those who care about the future of the world's forests.

£15.95 paperback ISBN 1 85383 378 9
£35.00 hardback ISBN 1 85383 379 7 224pp

Bad Harvest: The Timber Trade and the Degradation of the World's Forests
By Nigel Dudley, Jean-Paul Jeanrenaud and Francis Sullivan

Analyses the environmental effects the timber trade is having on forests – boreal, temperate and tropical – and considers the policies necessary to limit the damage. The authors discuss the changing nature of the trade and its effects on people working and living in the forests; the intensification of forest management; and the harmful effects of the pulp and paper manufacturing processes outside the forest. They conclude with a strategy for the timber trade which would help to conserve the forests at the same time as being economically viable.

£12.95 ISBN 1 85383 188 3 224pp Published in association with WWF–UK

The Economics of the Tropical Timber Trade
By Edward Barbier, Joanne Burgess, Joshua Bishop and Bruce Aylward

The first full analysis of the economic linkages between the tropical timber trade and forest management showing how appropriate policies and market incentives can lead to sustainable forestry.

£14.95 ISBN 1 85383 219 7 224pp

Controlling Tropical Deforestation
By Alan Grainger

Comprehensive introduction to the problem of tropical deforestation giving a fresh analysis of the causes and the policies needed to slow and prevent it.

'*Well produced and very readable... will be widely quoted in the years ahead*' Scottish Geographical

£14.95 ISBN 1 85383 142 5 310pp

The Earthscan Reader in Tropical Forestry
Edited by Simon Rietbergen

A collection of seminal contributions to the various aspects of the debate over how to conserve and manage the world's diminishing reserves of tropical forests.

£22.95 ISBN 1 85383 127 1 350pp